Cambridge Elements

Elements in Organizational Response to Climate Change

WHO TELLS YOUR STORY?

Women and Indigenous Peoples Advocacy at the UNFCCC

Takumi Shibaike
Syracuse University

Bi Zhao
Gonzaga University

Shaftesbury Road, Cambridge CB2 8EA, United Kingdom

One Liberty Plaza, 20th Floor, New York, NY 10006, USA

477 Williamstown Road, Port Melbourne, VIC 3207, Australia

314–321, 3rd Floor, Plot 3, Splendor Forum, Jasola District Centre,
New Delhi – 110025, India

103 Penang Road, #05–06/07, Visioncrest Commercial, Singapore 238467

Cambridge University Press is part of Cambridge University Press & Assessment,
a department of the University of Cambridge.

We share the University's mission to contribute to society through the pursuit of
education, learning and research at the highest international levels of excellence.

www.cambridge.org
Information on this title: www.cambridge.org/9781009472937

DOI: 10.1017/9781009472920

When citing this work, please include a reference to the DOI 10.1017/9781009472920

First published 2025

A catalogue record for this publication is available from the British Library

ISBN 978-1-009-47293-7 Hardback
ISBN 978-1-009-47289-0 Paperback
ISSN 2753-9342 (online)
ISSN 2753-9334 (print)

Who Tells Your Story?

Women and Indigenous Peoples Advocacy at the UNFCCC

Elements in Organizational Response to Climate Change

DOI: 10.1017/9781009472920
First published online: January 2025

Takumi Shibaike
Syracuse University

Bi Zhao
Gonzaga University

Author for correspondence: Takumi Shibaike, tshibaik@syr.edu

Abstract: Thousands of civil society organizations (CSOs) attend the Conferences of the Parties (COPs) of the United Nations Framework Convention on Climate Change (UNFCCC) every year. Through their advocacy work, CSOs define and redefine what "climate change" is really about. This Element focuses on climate advocacy for women and Indigenous peoples (IPs), two prominent climate justice frames at the UNFCCC. Which CSOs advocate for women and IPs? How and why do CSOs adopt gender and Indigenous framing? Bridging the literature on framing strategy and organizational ecology, it presents two mechanisms by which CSOs adopt climate justice frames: self-representation and surrogate-representation. This Element demonstrates that, while gender advocacy is developed primarily by women's CSOs, IPs advocacy is developed by a variety of CSOs beyond IPs organizations. It suggests that these different patterns of frame development may have long-term consequences for how we think about climate change in relation to gender and IPs.

Keywords: CSO, climate justice, UNFCCC, transnational advocacy, organizational ecology

ISBNs: 9781009472937 (HB), 9781009472890 (PB), 9781009472920 (OC)
ISSNs: 2753-9342 (online), 2753-9334 (print)

Contents

1 Introduction 1

2 Theoretical Framework 9

3 The Analysis of CSO Framing Efforts on Social Media 22

4 The Analysis of Women and Indigenous Peoples Advocacy 44

5 Conclusion 62

Appendix 67

References 80

1 Introduction

1.1 Overview

Since its establishment in 1992, the United Nations Framework Convention on Climate Change (UNFCCC) has stipulated and governed the multilateral negotiations on climate change. In recent years, while the negotiations are making little progress on the actual greenhouse gas reduction, as the national governments lock heads with one another, a different trend has been on a steady rise. Climate activism, especially with its focus on climate justice, has become a highly visible presence and prominent voice at the annual Conferences of the Parties (COPs) under the UNFCCC.

In 2017, Fiji became the first small-island state to host the COP. Although it was eventually located in Bonn, Germany, COP23 witnessed an increase of presence and events from small-islanders and indigenous peoples (IPs) who called for attention and specific actions on climate justice. At one event, then Prime Minister of Fiji and President of COP23, Frank Bainimara, spoke about the disproportional hardship for small islands, in front of a colorful background of tropical flowers, starkly contrasting a black-and-white sign that says "Ban Fossil Fuel!" Many other local and IPs organizations, with their vibrant dance, songs, and pictures, vividly depicted how they suffered from the tremendous, irreversible consequences of climate change that had almost wiped out their communities and lands off the map.

At the same COP, women's groups also took the stage with loud and clear voices. Whether it was a chant-and-dance at the Blue Zone, a silent protest by wearing mustache to name-and-shame the patriarchal practices embedded in the governing system, or events featuring local women's leadership in climate solutions, they demanded more gender-specific data related to climate impact and called out the gender gap and the lack of gender-sensitivity in climate policymaking.

These voices, often championed by civil society organizations (CSOs), have clearly entered the UNFCCC process. The multilateral process also responded to their demands. In the same year, COP23 recorded the adoption of two important documents: the Local Communities and Indigenous Peoples Platform, which supports the exchange of local and indigenous knowledge regarding mitigation and adaptation,[1] and the Gender Action Plan, which promotes gender equality in climate actions.[2] The CSOs championing climate justice never stopped. Although CSOs are observers (i.e. not Parties to the UNFCCC), they

[1] https://unfccc.int/resource/docs/2017/sbsta/eng/06.pdf (Accessed: October 20, 2023).

[2] https://unfccc.int/resource/docs/2017/sbi/eng/l29.pdf (Accessed: October 20, 2023).

continued to bring their issues to the annual COPs. Whether it is Side Events inside the venue or activism and protests outside, they have garnered growing attention from national delegates, international organizations, and the general public.

As non-Party observers of the UNFCCC, these CSOs do not have formal political power. However, they continue to define and redefine the meaning and narratives about climate change through issue framing, a strategy to highlight certain aspects of an issue in order to draw public attention (Albin, 1999; Allan and Hadden, 2017; Snow et al., 1986). In the early years of the UNFCCC, climate change advocacy focused primarily on scientific frames, such as greenhouse gas emissions and technical solutions to reduce them (Bäckstrand and Lövbrand, 2006). Over the past 30 years, the number of CSO participants increased significantly, and so did the diversity of CSOs beyond environmental CSOs. Today, climate change is no longer just about science. Many CSOs frame climate change as a social justice issue, such as gender inequality and IPs' rights violation.

This study examines different ways in which CSOs advocate for climate justice at UNFCCC COPs. We focus on two climate justice frames that have become prominent in recent years: a gender-climate frame and an indigenous-climate frame (Bäckstrand and Lövbrand, 2019). In so doing, we ask what kinds of CSOs develop gender and indigenous frames and why they chose theses frames over other climate frames. To answer these questions, we draw on the theoretical frameworks of framing strategy and organizational ecology. By connecting climate frames with the resource space surrounding CSOs, we argue that CSOs' framing efforts are shaped by both the availability of resources and the expertise of individual organizations. In turn, these framing efforts affect interorganizational dynamics and the evolution of climate frames over time. In particular, we propose two mechanisms of frame development based on interest representation: *self-representation* and *surrogate-representation*.

We find that women's CSOs are self-representative in that they lead and shape advocacy that highlights the need of addressing gender equality in global climate governance. By contrast, IPs advocacy takes on both self-representation and surrogate-representation with the latter growing prominently. In surrogate-representation, a diverse group of CSOs, including non-indigenous CSOs specializing in issues like forestry and agriculture, advocate for IPs. We argue that the patterns of frame development depend on the willingness of early-comer CSOs to discipline their advocacy narratives. A core group of early-comer CSOs leads and shapes the narratives about gender-related issues in climate governance, whereas IPs organizations do not exercise such power but embrace diverse voices. We suggest that these different patterns can have a long-term impact on how we think about the situations surrounding women and IPs in relation to climate change.

1.2 Contributions

This study sheds new light on the role of CSOs in global climate governance, especially the ones that are specialized in particular issues or people groups. These CSOs tend to be small and lesser known groups, but they are nevertheless experts in their areas of specialization. While these CSOs are often treated as a trivial category of "followers" or "free-riders" (Bob, 2005; Murdie, 2014), recent research on CSOs in global governance has shown how such specialist CSOs might shape the institutional environment surrounding themselves (Bush and Hadden, 2019; Eilstrup-Sangiovanni, 2019; Shibaike, 2023). By highlighting two advocacy communities – gender and IPs – at the UNFCCC, we show how specialization can lead to remarkably different organizational dynamics in climate advocacy. Doing so contributes to a better understanding of how CSOs as a whole carry out climate advocacy and shape the narratives about climate change at the UNFCCC.

This study also contributes to our understanding of IPs' participation in global climate governance. Existing research has shown that IPs have experienced political, economic, and epistemological barriers, as they attempt to engage with the UNFCCC process (Comberti et al., 2019). This study seeks to move beyond the issue of marginalization and highlight the ways in which IPs developed their own voices. We hope to highlight the contributions as well as struggles of IPs at the UNFCCC and suggest more engagement and future research on this topic.

Finally, this study lends insights into the question of democratic legitimacy in global governance (Bexell, Tallberg, and Uhlin, 2010; Steffek, Kissling, and Nanz, 2007; Zhao, 2023). The UNFCCC process is generally considered as a relatively open space where different stakeholders in climate governance can join. It is known for its accessibility for non-Party observer organizations, and thousands of CSOs have flooded to the annual COPs over the years. However, attendance does not equate meaningful and engaging participation. We shift our focus away from mere institutional access and closely examine the mechanism of framing efforts at the UNFCCC. In so doing, we show power dynamics among CSOs and their long-term effects on meaning-making around climate change and participation patterns.

Next, we first offer historical trajectories of women's CSOs and IPs participation in the UNFCCC process, including the institutional responses and arrangements related to their advocacy efforts. We then introduce an outline of the Element.

1.3 Gender and Indigenous Advocacy at the UNFCCC

1.3.1 Women and Gender Advocacy

The linkage between women and climate change came forth during the 1992 UN Conference on Environment and Development at Rio de Janeiro, where the frame connecting women's rights and sustainable development first emerged (Friedman, 2003). After the UNFCCC was established in 1992, however, the regime had little to say about gender equality. From research scientists to national delegates, men have constituted the majority of the research and policymaking forces (Nagel, 2015). The demand for awareness of the linkage between women and climate change started to surface in the early 2000s. In 2001, gender concerns were officially brought into the UNFCCC process, in which the Secretariat was tasked with examining the gender composition at the UN climate governance. However, no mechanism was adopted to ensure gender quality. The failure of Copenhagen, the rising doubt about the UN setup, and the fact that the system was dominated by men, indicated that it was time to call for the participation of women in global climate governance (Buckingham, 2010).

The endeavor to organize movements around women's rights in climate governance came into public view in 2007 during the COP13 meeting in Bali, Indonesia. The Women's Caucus was founded, which later gained the official UNFCCC constituency status as the Women and Gender Constituency (WGC). The initial women and gender CSOs included Women in Europe for a Common Future (WECF), Gender Climate Change, Women's Environment and Development Organization, and Life.e.V. They, along with other CSOs, demanded the UNFCCC be a "gender-sensitive regime." At the same time, they started "holding joint meetings, lobbying delegates, organizing side events, participating in protests, and advocating for the inclusion of gender-specific text in the negotiations" (Ciplet, 2014: 81). Despite the short-lived attention for gender equality in Indonesia, the COP13 established a fertile ground for the participation of influential women's CSOs in climate governance. For example, in 2008, GenderCC – Women for Climate Justice established a clear linkage between climate change and gender inequality and called for fundamental changes beyond including women in the climate change establishment.

The WGC was formally established in 2009. In the same year, members of the WGC lobbied at the COP15 in Copenhagen for the inclusion of a shared vision "Preamble to the Convention" with the "full integration of gender perspectives" (Ciplet, 2014: 82). However, despite CSOs' persistent advocacy, the Copenhagen Accord did not contain any reference to women or any gender-sensitive language. Overall, the institutional inclusion and engagement of women at the UNFCCC did not gain real momentum between 2009 and 2011, except for a few gender-specific provisions added in the Cancun Agreement in 2010 and reiterated in the Durban Platform in 2011. However, these commitments remained vague and more of a lip service than a sincere policy change, unable to influence any practices at the national level.

Finally, in 2012, at the COP18 meeting in Doha, Qatar, the mobilization of women's CSOs started paying off. The decision adopted at COP18 included the language of "promoting gender balance and improving the participation of women in UNFCCC negotiations and in the representation of Parties in bodies established pursuant to the Convention or the Kyoto Protocol" (UNFCCC, 2012). The emphasis on the importance of women negotiators in the formal text was heralded by many as the "Doha Miracle." The decision reflected an official commitment to include women in the UNFCCC process and required that data on gender participation be collected and made public, which would make research on women's representation much easier than before.

The following year, the UNFCCC again recognized the need for gender balance in its decision-making processes. As a result, expert bodies of the UNFCCC directed the International Panel on Climate Change (IPCC) to include the language of gender in its 2013 report. Although women's representation was formally mandated, representatives of CSOs claimed that a mere numerical increase had not improved the quality of women's participation. The initiative to increase women delegation is only a "tick-box" approach: the rising number of female delegates cannot be translated into gender equality in climate governance. Mandating women's representation was the first and vital step (Alston, 2015).

The WGC did not stop after the inclusion of gender-specific clauses into the formal texts in 2012 and 2013. They kept advocating for the cause in the years that followed. At the COP20 in Lima, Peru in 2014, the WGC argued that "the COP parties were failing to implement solutions that considered the critical role of women and the importance of gender equality in tackling climate change" (Kuyper and Bäckstrand, 2016: 62). After the 2015 Paris Agreement was adopted at the COP21, the WGC was concerned about how to implement the hard-fought languages of human rights, gender equality, and other principles included in the preamble of the Paris Agreement into practical steps and policy instruments. The WGC successfully persuaded the Parties during the COP22 in 2016 to develop a "Gender Action Plan (UNFCCC, 2017, Decision 3/CP.23)" as an implementation roadmap for them. The Plan was a milestone that underlined the urgency of integrating gender into climate-related policymaking. The goal was to ensure that women have a say in climate change decision-making and that the interests of all genders are sufficiently represented at the UNFCCC.

1.3.2 Indigenous Peoples Movement

IPs began attending the UNFCCC COPs in 1998. The first groups of IPs at the COP4 in Buenos Aires came from North America who issued "A Call to Action: The Albuquerque Declaration" (1998). Over the years, IPs' presence increased, and several institutional platforms have formed to gather IPs coming to the UNFCCC process. In response to the growing IPs' presence at the COP7 in 2001, the UNFCCC recognized the IPs as one of the non-state observer constituencies, known as the Indigenous Peoples Organizations (IPO) (Belfer et al., 2019). Several years later, in 2008, the International Indigenous Peoples' Forum on Climate Change (IIPFCC) was established as the caucus for IPs and their allies in the UNFCCC process.[3] Its mandate was to agree specifically on what IPs will be negotiating for in the UNFCCC process, hence providing a channel to develop unified positions for the IPs coming to the venue.

The development of IPs advocacy inside the UNFCCC has drawn heavily from the international legal framework developed outside of the UNFCCC. For instance, the UN Permanent Forum on Indigenous Issues was established in 2000 to serve as a high-level advisory body to the UN Economic and Social Council.[4] Its composition – a mixture of nation-states and IPs – has become a template for developing proposals for advisory bodies at the UNFCCC (López-Rivera, 2023). One important legal framework developed outside of the UNFCCC is the Declaration on the Rights of Indigenous Peoples (UNDRIP) adopted in 2007. The UNDRIP positions the right to self-determination and collective rights to lands, territories and resources at its core principle. It addresses the relationships between indigenous peoples and their land and resources, and how such relationships are essential to their identity and well-being.[5]

Parallel to the UNDRIP, the REDD (Reduction of Emissions from Deforestation and forest Degradation) program was introduced in 2007. It has been a controversial decision among IPs because of the risk of dispossession of IPs' lands (Claeys and Delgado Pugley, 2017: 3). A range of reactions spawn from

[3] International Indigenous Peoples' Forum on Climate Change. www.iipfcc.org/who-are-we (Accessed: June 13, 2023).

[4] https://www.un.org/development/desa/indigenouspeoples/about-us/permanent-forum-on-indigenous-issues.html (Accessed: June 13, 2023).

[5] https://www.un.org/development/desa/indigenouspeoples/wp-content/uploads/sites/19/2018/11/UNDRIP_E_ web.pdf (Accessed: August 22, 2023).

the IPs community, ranging from explicit opposition to participation in the program in exchange for cash and a point of contact with the broader mitigation effort. At the same time, many of the IPs' organizations sought to integrate the UNDRIP into the UNFCCC mitigation mechanism. They launched a No Rights, No REDD campaign, as many UNFCCC Parties refused to make an explicit mention of the UNDRIP (Claeys and Delgado Pugley, 2017: 4). IPs expressed disappointment at the lack of political will by Parties to include any reference to the language of indigenous rights crafted under the UNDRIP. The IPs also saw the COP15 in Copenhagen as an opportunity to reverse this political stalemate; however, the COP15 turned out to be a complete disaster for CSOs, as it severely restricted CSO participation in response to disruptive protest events. But the event became an awakening point for CSOs to reconsider the governance of climate change and triggered the climate justice movement worldwide. It prompted the People's Climate Summit in 2010 in Bolivia, where representatives of IPs voiced their views on the management of indigenous territories, extractive industries, and REDD. The consensus among the IPs community was to support the process of recognition of the UNDRIP (Claeys and Delgado Pugley, 2017). They were successful in this effort as the UNDRIP was noted as a safeguard provision of the REDD mechanism in 2010 (Claeys and Delgado Pugley, 2017: 5). However, in the following years, no concrete action was taken based on the Safeguard Information System, which was supposed to guide international reporting on how countries address and respect safeguard standards (Claeys and Delgado Pugley, 2017: 6).

While the IPs' voices and the REDD have been entangled for years without a clear resolution, the recognition and inclusion of indigenous knowledge in the Paris Agreement was a notable victory. More specifically, the Paris Agreement encourages Parties to consider "the best available science" along with "traditional knowledge, knowledge of Indigenous peoples and local knowledge systems" (UNFCCC, 2017: Article 7.5). The accompanying decision to the Paris Agreement introduced an institutional innovation that materialized this formal recognition by establishing the Local Communities and Indigenous Peoples Platform (LCIPP).[6] It was a unique achievement, as it established the first formal, permanent, and distinct space created for IPs within the UNFCCC. Unlike existing initiatives, the LCIPP was designed to feed directly into the negotiation process as a working group. It is a forum for fostering dialogue between Parties and IPs about what inclusive national and international climate

[6] Report of the Conference of the Parties on Its Twenty-First Session, decision 1/CP.21, para 135, UN Doc. FCCC/CP/2015/10/Add.1.

actions should look like, while increasing the capacity of IPs to implement their own projects (Belfer et al., 2019; López-Rivera, 2023).

Despite its importance, the LCIPP created under the Paris Agreement has been an inadequate mechanism for IPs participation in the UNFCCC process, as it fails to acknowledge IPs' participation as inherent rights and avoids recognizing the colonial systems and practices that marginalize IPs (Maldonado et al., 2016; Raffel, 2016; Tormos-Aponte, 2021). It also left much ambiguity for interpretation when Parties seek implementation based on the LCIPP. Following the COP21 in Paris, the interpretation and design of the LCIPP have been in disarray due to its vague language, allowing for various interpretations and submissions from Parties regarding how to implement Paragraph 135, which called for strengthening indigenous knowledge in climate actions. Some Parties, such as Canada and Ecuador, become the champions for IPs, while others have not (for example, Brazil only called for a website to "deposit" indigenous knowledge) (López-Rivera, 2023). Therefore, from the IIPFCC to the REDD+ and then to the LCIPP, the right to self-determination and the right to natural resources on indigenous lands and territories remain insufficiently fulfilled.

1.4 Outline of the Element

Against this background of gender and IPs' advocacy at the UNFCCC, this study examines how CSOs working on these two areas built their narratives through framing efforts. To explore CSOs' framing efforts, we begin by presenting our theoretical framework in Section 2. We connect the literature on advocacy framing with the theory of organizational ecology to explain the observed differences in framing processes, outcomes, and future implications. In particular, we explain how self-representation and surrogate-representation may explain the process of frame development at the UNFCCC process.

Our empirical research adopts a mixed-method approach. In Section 3, we use Twitter (now X) data to empirically observe variations in the use of different frames among CSOs during COP21 in Paris, a milestone event in global climate governance. Twitter is (or was during COP21) a widely used social media platform among CSOs. It provides a channel for them to publicize their voices with a minimum entry barrier. We first show the landscape of civil society advocacy at the UNFCCC based on computational text analysis. We then use statistical models to analyze the characteristics of CSOs associated with gender and IPs advocacy. The findings illustrate the patterns of self-representation and surrogate-representation in gender and IPs advocacy.

In Section 4, we draw on the in-depth interviews conducted at COP23, COP24, and COP27 to qualitatively explore CSOs' framing efforts for gender and IPs. We leverage accounts from CSO representatives to show how CSOs perceived their institutional environment, how they interact with one another at the UNFCCC, and their long-term impacts on both climate advocacy and the organizations involved.

Section 5 concludes and considers the implications for future climate advocacy and research agenda. The theory and data that we provide in this Element suggest new ways to look at the process of frame development among CSOs. We call for further attention to organizational population dynamics in analyzing advocacy frames. We end with practical considerations for the future of climate advocacy based on the findings in this Element.

2 Theoretical Framework

During the past few decades, civil society organizations (CSOs) have became a group of salient political actors in the governance of transnational issues, including human rights, development, and the environment and climate change. They have challenged the state-centric system of global governance by exercising their normative and symbolic power (Keck and Sikkink, 1998). CSOs play a crucial role at multiple stages of international policy-making processes, including norm creation, agenda-setting, policy formation, monitoring, and enforcement of international agreements (Steffek, 2013; Tallberg, Sommerer, and Squatrito, 2013). Today they are widely regarded as an indispensable part of world politics. The theorizing of CSOs progressed with the emergence of constructivist theory in the field of international relations (Finnemore and Sikkink, 1998). With its emphasis on social and ideational factors in world politics (Finnemore and Sikkink, 2001), a constructivist framework became a major approach in explaining how CSOs, which lack the conventional forms of material power, can effect political and normative changes (Willetts, 2001). Over time, scholarly questions about CSOs and their effects on world politics expanded. Scholars ask not only whether or not CSOs matter in world politics, but why and under what conditions they matter. Much academic attention has been paid to the strategies of CSOs, such as building connections through transnational networks (Allan and Hadden, 2017; Keck and Sikkink, 1998; Shawki, 2011), naming and shaming (Dietrich and Murdie, 2017; Hafner-Burton, 2008; Hendrix and Wong, 2013), and the combination of insider and outsider strategies at intergovernmental organizations (Fox and Brown, 1998; Hadden, 2015).

We build on the existing efforts to analyze CSO strategies. This section focuses on how and why CSOs choose their frames at the UNFCCC and how their framing choices in turn affect inter-organizational dynamics. First, we discuss the literature on advocacy framing to identify the need to examine the origin of advocacy frames. We then connect framing strategies with the theory of organizational ecology and posit that framing is a way to extract social and economic resources from their resource base (i.e. niche). Next, we advance our understanding of CSO advocacy at the UNFCCC by presenting two different mechanisms by which CSOs adopt a particular frame and represent the interest of people groups, namely, *self-representation* and *surrogate-representation*. In the first mechanism of self-representation, CSOs align their frames with their inherent identity and mission and claim the exclusive ownership of their frame. In the second mechanism of surrogate-representation, CSOs develop advocacy discourse for "others" that are outside of their direct constituency. Finally, we look at the temporal processes of framing choices. We argue that the strategies of early-comer CSOs that initially developed novel climate frames have far-reaching consequences for how follower CSOs adopt climate frames. While some CSOs attempt to *discipline* the narrative around a particular frame, others do not, resulting in organizational populations that look remarkably different within the UNFCCC space. Our argument highlights the importance of small and specialist CSOs in the UNFCCC, departing from an overwhelming emphasis on the power of large CSOs as gatekeepers.

2.1 Advocacy Framing among CSOs

Goffman (1974: 21) defines frames as "schemata of interpretation" that enable actors to "locate, perceive, identify and label" specific events and occurrences. Framing is, in essence, a process to make sense of the world with these structures of interpretation. Similarly, Gamson and Modigliani (1987: 143) refer to frames as a "central organizing idea or storyline," which uses interpretative categories to explain and describe complex processes. As such, framing processes are characterized by interpretative and selective nature. Frames and the framing processes allow both framers and audiences to incorporate certain interpretations of phenomena. Through this meaning-making process, framing guides one's decisions of what is meaningful and desirable. Because framing is interpretative, the framing process communicates and highlights certain elements of an issue while deflecting away from other aspects that may be less relevant (Snow et al., 1986).

We take the position that framers and their agency are central to the process of framing, as opposed to the post-structuralist account that they are produced by climate governmentality (Bäckstrand and Lövbrand, 2006; Dowling, 2010). Framers are not passive recipients of frames but are engaging in ongoing processes of meaning-making (Gamson, Modigliani et al., 1987; Snow et al.,

1986). Snow and Byrd (2007) emphasize the agency of actors who engage in framing efforts. They argue that actors are not just promoters of certain ideas. Rather, framing indicates that the "signifying agents actively [engage] in producing and maintaining meaning for constituents, antagonists, and bystanders" (Snow and Byrd, 2007: 123). If framing proves to be successful, the audience will think about an issue in the way that it is framed, focusing on certain elements over others (Nisbet and Scheufele, 2009). As such, framing is a tool of influence for CSOs, which typically lack material sources of power. Various studies confirm that framing plays a crucial role in organizing and advancing collective action (Benford and Snow, 2000; Boscarino, 2016; Keck and Sikkink, 1998; Tarrow, 2005).

Analyzing CSOs working in global climate governance, Allan and Hadden (2017: 601) show how framing expanded civil society coalitions and increased the issue salience of loss and damage at the UNFCCC. They point out that justice-based issue framing "increased attention to the issue and laid the groundwork for the formation of new coalitions, increasing solidarity among NGOs and state actors." Schapper, Wallbott, and Glaab (2023) highlight the "community of practice" in climate justice advocacy and find the moderation of claims among various actors. Others see climate-bandwagoning as a form of framing (Hjerpe and Buhr, 2014; Jinnah, 2011), where "framers establish strategic links to expand their mission scope by including new-climate related goals (Jinnah, 2011: 4)," often linking climate change to non-climate issues (Hjerpe and Buhr, 2014).

While existing research on CSOs' framing efforts often focuses on the outcome, such as coalition building and increased issue salience, we know little about who produced such a variety of climate frames in the first place. As we detail next, the composition of climate frames at the UNFCCC is a product of both CSOs' moral commitments and incentives for organizational survival. This is an important question because climate change encompasses a wide range of frames from which actors adopt in their advocacy work. Climate framing has been continuously expanding its scope. During the first decade of the UNFCCC, the issue was discussed very much within scientific framing. However, since COP15 in Copenhagen, civil society activism has shifted toward framing that highlights social and ethical concerns of climate change, such as social justice, development, and human rights (Hadden, 2015). For example, Hjerpe and Buhr (2014: 105) identified eight different frames that feature various aspects of climate change and its consequences (science, social progress, economic, public accountability, moral or ethics, uncertainty, Pandora's box, and conflict).

We argue that CSOs develop advocacy frames both for their commitment to the cause *and* for their organizational survival. Our study contributes to this effort by carefully unpacking, tracing, and understanding the driving forces of climate frames, and how their interactions may eventually shape the characteristics of CSO populations at the UNFCCC. In particular, we provide two processes of frame development by which the interest of climate-affected groups is represented (self-representation and surrogate-representation). Our argument has important implications for policy debate and agenda-setting research, as framing offers profound consequences for structuring policy debate as well as shaping public opinion and media attention around the issue (Boscarino, 2016).

2.2 Organizational Ecology, Niche Development, and CSO Framing

To better understand CSOs' framing efforts, we employ and extend the theoretical framework of organizational ecology and integrate it with the literature on advocacy framing. We argue that CSOs use frames and narratives to extract social and material resources. While existing literature has focused on advocacy framing as a tool to further the cause of CSOs, we contend that framing efforts are for the benefit of CSOs as well. Our argument contributes to the long-standing debate around CSOs, whether they are motivated by principled values or organizational self-interest (Prakash and Gugerty, 2010). By theorizing how CSOs develop advocacy frames (rather than identifying their effects), we demonstrate that they can be principled and self-interested actors at the same time. In what follows, we first layout the fundamental concepts in organizational ecology and then connect them with CSO framing strategies.

2.2.1 Organizational Ecology and Niche

The theory of organizational ecology was developed in the field of sociology to explain the diversity of organizations in a given market (Hannan and Freeman, 1977). In this theoretical framework, a similar type of organizations (e.g. private organizations as opposed to intergovernmental organizations) are treated as an organizational *population*. Different populations depend on different resources for organizational survival, just like different species require different sets of resources for subsistence in an ecosystem. The set of resources on which a given organizational population subsists is referred to as a *niche* (Freeman and Hannan, 1983).

The concept of a niche is important to our understanding of change and survivability of organizations. The traditional notion of a niche typically emphasizes the availability of material and economic resources for organizational survival (Hildebrandt and Chua, 2017). We expand the concept of a niche and include both material and *social* resources. A niche can be material in the sense that physical assets, such as land, buildings, and raw materials, provide functional requirements for organizations (Baum and Oliver, 1991; Hannan and Freeman, 1977). A niche can also be social, as non-material support from a variety of actors, including peer organizations, allows organizations to survive. Their survivability does not rely exclusively on financial resources because many CSOs operate on a small budget. In this sense, well-connected organizations in advocacy networks, often referred to as "gatekeepers" (Carpenter, 2007), have abundant social resources. CSOs can also leverage a stock of social and moral support for particular ideas and norms, which provide them with legitimacy and authority (Stroup and Wong, 2017). The abundance and poverty of those resources determine the rise and fall of organizational populations.

2.2.2 Niche Development and Specialization

As any resource, including social one, is finite, competition among organizations for niche occupancy is inevitable. In the early stage of niche formation, a niche is populated by a small number of organizations (Abbott, Green, and Keohane, 2016; Hannan and Freeman, 1989). That is, the *density* of an organizational population is low, and competition among organizations is relatively moderate. However, once the number of organizations in the niche begins to grow, an increasing density of organizations intensifies market competition. This is reflected in framing strategies as well. For example, in environmental policymaking, CSOs develop competing frames and seek to establish rhetorical control over the terms of policy debate (Boscarino, 2016). As organizations begin to experience a lack of material and social resources for survival, the growth of an organizational population will be severed. This competition will eventually lead to an inflection point, where a negative relationship starts to display between population density and the growth rate of organizations within the niche.

Once the growth of an organizational population slows down, some organizations start targeting a narrower niche than existing organizational populations do. Organizational ecologists call such organizations as *specialist* organizations. Specialist organizations are able to survive in the same market with *generalist* organizations by focusing on a particular set of social and material resources for subsistence (Carroll, 1985). For example, specialist conservation CSOs like Save the Rhino International can co-exist with generalist conservation CSOs like the Wildlife Fund for Nature (WWF) in the same market by appealing to a particular set of resources (Bush and Hadden, 2019; Shibaike, 2023).

When the study of global governance adopted organizational ecology, many studies focused on the rise and fall of organizational populations (Abbott, Green, and Keohane, 2016; Eilstrup-Sangiovanni, 2021; Lake, 2021). For example, Bush and Hadden (2019) demonstrate that the rate of NGO founding declines as the sector becomes densely populated. Others focused on the strategies of niche exploitation, namely specialism and generalism, among CSOs (Eilstrup-Sangiovanni, 2019; Shibaike, 2023). Generalist CSOs tend to be more materially well-resourced than specialist CSOs, as the former has access to a wide range of supporters, including governments, corporations, and the general public.

However, the size of material resources that CSOs enjoy does not necessarily translate into their abilities to shape narratives around them. As Stroup and Wong (2017) argue, large CSOs tend to take moderate strategies that are unlikely to expand or curve out a new niche. By contrast, specialist CSOs may be supported by highly engaging supporters who can bring in more resources into their niche (Shibaike, 2023). We advance existing insight by demonstrating that there are different ways in which organizations *specialize* into particular niches.

2.2.3 CSO Framing as Niche Development

Connecting organizational specialization with CSOs' framing choices, we argue that CSOs use frames and narratives to extract social and economic resources within their niche. In line with Eilstrup-Sangiovanni's (2019: 381) work on an "advocacy niche," we conceptualize the niche as a set of material and social resources that include "dominant ideas, norms, and cognitive frames relevant to an issue, information about policy preferences and choices." While Eilstrup-Sangiovanni's work highlights the niche created by technological advances and legal changes, we focus on the fact that CSOs can expand their niche by their own advocacy work. That is, the fluidity and variability of climate discourse are produced by CSOs' framing efforts themselves. A stock of social and moral support for a particular frame becomes a unique type of resources that can help sustain the lives of CSOs. Thus, there is an interdependent relationship between the abundance of resources (or lack thereof) within a given niche and the CSOs that subsist on it.

In the context of climate advocacy, CSOs make framing efforts in order to guide their audiences to understand climate change from a particular angle and to extract economic and social resources for organizational survival. The complex nature of climate change provides a large number of framing options, and we observe a wide variation in the frames that CSOs use on the ground. That is, the meaning of climate change is constituted by a collection of various frames that emphasize different elements of climate change (Hjerpe and Buhr, 2014; Nisbet and Scheufele, 2009; O'Neill et al., 2010). As such, frames for climate change – climate frames – are empirically fluid and variable, and hence less restrictive in terms of who can engage with a certain climate frame.

The UNFCCC process, in which nation-states gather annually to negotiate greenhouse gas emission reduction goals and other climate-related issues, provides an ideal venue for tracing the evolution of climate frames. Traditionally, climate change was framed as a scientific problem. Scientific experts acquired technologies to monitor and provide evidence of global environmental degradation, which formed the basis of climate change advocacy from the period leading up to the establishment of UNFCCC in 1992 (Bäckstrand and Lövbrand, 2006). This scientific frame has continued to occupy mainstream discourse. In the past decade, however, climate advocacy began to emphasize alternative frames, especially those highlighting social and ethical concerns about climate change (Allan and Hadden, 2017; Hadden, 2015). Frames emphasizing climate justice, rights-based approaches, development, and marginalized groups, such as women and IPs, are now widely accepted among CSOs.

Empirically, a niche for CSOs at the UNFCCC corresponds to the advocacy frames since CSOs adopt frames to garner material and social resources for their advocacy efforts. Such frames include mitigation, adaptation, climate justice, gender equality, and the rights of indigenous peoples, to name a few. These frames also define the key characteristics of supporters who provide resources for organizational survival. By focusing on the frames that constitute the relationship between the niche and CSOs, we distinguish the ecology of climate CSOs and the ecology of the climate frames that they construct. That is, the presence of CSO representing specific interests, such as rights of people groups and solutions to specific issues, does not automatically translate to a clear and loud voice for those interests. Importantly for CSOs, organizational identities, often defined in their mission statements or membership in a UNFCCC constituency,[7] do not necessarily correspond to the *behaviors* of CSOs.

Therefore, while CSOs at UNFCCC can be defined in terms of the characteristics of their missions, those attributes do not tell us about their framing efforts on the ground. We motivate our analysis of CSOs with the idea of organizational specialization, but we depart from existing research by focusing on different ways in which CSOs specialize framing efforts at the UNFCCC.

[7] A constituency is a voluntary network of CSOs to share information and assist each other for participation in the UNFCCC process. There are currently nine official constituencies.

While recognizing that organizational ecology is primarily a structural theory, we emphasize that specialist CSOs have the power to shape the institutional environment by themselves. In so doing, we demonstrate that specialist CSOs develop their frames differently. That is, specialization is not a uniform phenomenon but has a variation within it. In particular, in the process of finding their own niche of specialization, CSOs do not always conform to the sectors they nominally belong to. For example, a CSO in the public health sector may take on narratives around gender equality, or a CSO works on forest conservation can develop discourse and understanding about the local IPs who have a close connection to forests. In the following sections, we detail how populations of CSOs can leverage particular climate frames to formulate and shape specialization.

2.3 Framing Mechanisms: Self-Representation and Surrogate-Representation

We argue that the mechanisms of frame development can be understood as the mechanisms of interest representation. We propose two major mechanisms here: *self-representation* and *surrogate-representation*. Conceptually, self-representation is characterized by the direct accountability relationship between specialist CSOs and their supporters who share the same identity, in which principal-agent relations typically emerge. For instance, although women's CSOs are supported by both women and men, women tend to be the main principal of women's CSOs. Moreover, organizational missions often focus on different ways by which women's rights and well-being can be realized. Empirically, we expect that the frame choices of specialist CSOs correspond to their own organizational identities, as existing research has often assumed (Allan, 2020; Cabré, 2011). For example, women's CSOs may use gender frames for self-representation. As such, they develop an ownership of gender frames and receive recognition as the authority of women and gender issues in relation to climate change. However, direct accountability is not the only form of political representation (Saward, 2006).

Drawing on Mansbridge's (2003) work on political representation, we use the term "surrogate" to highlight the absence of direct accountability relationship between groups (e.g. IPs) and the CSOs who represent their interest. Existing research shows that the agency of IPs in climate governance has been obscured due to the lack of participatory mechanisms for IPs (Schroeder, 2010). As a result, the role of allyship with IPs has attracted much attention (Allan, 2020; Bond and Dorsey, 2010). While CSOs' advocating for IPs can be seen as allyship, there is usually power imbalance between IPs and CSOs. For example,

Northern conservation CSOs may support for Southern IPs' participation in programs like REDD+ through capacity building. Northern CSOs are generally well-intended, but treating this relationship as an allyship would obscure the reasons behind contention among IPs regarding REDD+, where some IPs are concerned that their agency and autonomy would be coopted by neoliberal market forces. Empirically, surrogate CSOs may find issue linkage and tap onto the niche that does not correspond to the CSOs' organizational identities. For instance, Northern CSOs working on issues such as forest, agriculture, and wildlife conservation can construct a narrative about IPs. Importantly, these CSOs are not directly accountable to IPs but their Northern supporters,[8] and they do not share indigenous identities as organizations. This can be a strategy to attract supporters for their causes and secure funding (Hildebrandt and Chua, 2017: 640). As a result, an indigenous-climate frame may be represented not only by IPs themselves, but by other specialist CSOs. Sections 3 and 4 will elaborate the two mechanisms empirically.

Analyzing the UNFCCC COPs, we argue that women's and gender advocacy at the UNFCCC is a form of self-representation. Over the past two decades, women's CSOs have become increasingly prominent at the UNFCCC COPs. Different women's CSOs have different answers to the question, "What does gender have to do with climate change?" Those who work on natural disasters and gender equality highlight the fact that women are more likely to die in climate-related disasters and that they are more likely to suffer from sexual harassment afterward. Others with expertise in gender sensitivity in urban planning argue that the urban infrastructure without gender consideration often induces more carbon emissions. Hence, all of these CSOs contribute to the development of gender-climate framing based on their specialization, while expanding the gender-climate niche as a whole at the UNFCCC.

By contrast, we argue that IPs' advocacy is a form of surrogate-representation. For example, in many market-based, payment for environmental service (PES) programs, local landowners and farmers are paid to manage their natural resources with the goal to reduce total carbon emissions when tackling climate change. In recent years, there has been an increasing number of initiatives to include IPs and local communities in these programs. The CSOs that design and manage these PES programs often do not share any indigenous identity, but they can nevertheless construct a narrative about IPs by framing

---

[8] For example, Forest Trends, which supports IPs in forests, are funded primarily by Northern donors (www.forest-trends.org/who-we-are/financial-information/, May 26, 2024). This does not necessarily mean that the organization is appropriating IPs' identities. We use Forest Trends as an example of a complicated accountability relationship between organizations, supporters, and IPs.

PES as a benefit to IPs. They may build a rhetoric about the role of IPs in the PES programs, reasons for their inclusion, and the benefits for them. As such, these CSOs may become "surrogates" of the IPs' voices, and their voice may shape an indigenous-climate frame. In surrogate-representation, the popularity of a frame outside of one's own sector creates an incentive for the CSO to align their issue frames with a popular and prominent frame (Zeng, Dai, and Javed, 2019). Especially when the frame signifies a justice-oriented moral connotation, such surrogate representation can attain a sense of legitimacy and integrity for the CSOs.

2.4 Disciplining in Niche Development

To further explain how specialist CSOs occupy a climate-related niche, we also highlight the disciplinary process by which early-comer CSOs reinforce the contour of a frame. Similar to the idea of gatekeeping (Bob, 2005; Carpenter, 2007), disciplining can give early-comers an advantage by providing the power to control the narrative with their "own" frame. Unlike gatekeeping, however, the disciplinary process also involves capacity-building for late comers. Although late-comers may express frustration toward the disciplinary process, they adapt to the norms of the UNFCCC process by learning necessary languages and skills. Importantly, CSOs may or may not exercise this power. We argue that whether or not early-comers exercise such disciplinary power depends on the broader social structure in which niche development takes place.

We argue that an important aspect of climate frame development and niche occupation is temporality. It matters when frames are introduced, by whom, in what order, and how long they have been around (Eilstrup-Sangiovanni, 2021). In particular, the initial set of organizations that occupy a given niche have a long-term impact on whether and how other organizations join the niche. In other words, early-comers enjoy legitimacy, expertise, and political advantages that shape the participation and opportunities of late-comers in the same framing space (i.e. niche).

The early-comer benefits are formulated in three ways. First, early-comers can define what it means to have a legitimate form of CSOs. The studies of organizational ecology find that an increase in the number of organizations with a specific form tends to lower the entry cost of this form in the same market (Eilstrup-Sangiovanni, 2019; Minkoff, 1997). This is a diffusion effect by which the organizational form of early-comers becomes more acceptable to late-comers through the "logic of appropriateness" (Abbott, Green, and

Keohane, 2016: 52). Second, early-comers have the opportunity to accumulate knowledge and expertise, which leads them to craft a mature and well-versed narrative over time (North, 1990; Pierson, 2000). This learning process allows early-comers to develop narratives and frames most suitable for the audiences at the UNFCCC. Finally, early-comers have ample opportunity to interact with other political actors, communicating their ideas and policy positions through various channels. Once they establish a foothold in the UNFCCC process, they can discipline late-comers in ways that will help advance their political positions.

Importantly, early-comers may or may not use these advantages to discipline the narratives and advocacy tactics of CSOs within the same niche. If early-comers exercise disciplinary power over other CSOs, they will produce a unified voice with other occupants of the same niche, although such early-comers may face frustration and criticisms expressed by the late-comers going through the disciplinary process. By contrast, if early-comers do not exercise disciplinary power, a wide variety of CSOs will adopt the frame in ways that advance their interests. This may add uncertainty to the general understanding of the issue featured in the frame. That is, the variety of CSOs adopting the same frame could help mainstream the issue in global governance beyond the UNFCCC, but it could also mean that the frame is appropriated by other CSOs for their own sake without substantially improving the issue itself.

Whether or not early-comers exercise disciplinary power depends on the broader social structure in which they are embedded. In addition to early-comer advantages, robust international norms legitimize early-comers whose frames and narratives conform to such international norms. In our case, the mainstreaming of women and gender discourse in global governance helped early-comer CSOs to discipline the narratives about women and gender in climate change. Since the UN Fourth World Women's Conference in Beijing in 1995, more and more international institutions have adopted the language of women and gender in their policymaking processes (Hafner-Burton and Pollack, 2002). Although the progress has been slow in terms of substantive gender equality at the UNFCCC, women's CSOs were socialized into appropriate behaviors, including strategic framing, in global governance. By contrast, the recognition of IPs' rights has not permeated across international institutions. The UNDRIP was a major step forward, but participatory arrangements for IPs in global governance are still limited (Belfer et al., 2019; Zurba and Papadopoulos, 2023). The structural conditions surrounding IPs do not provide a strong guideline as to what appropriate behaviors should look like in IPs advocacy, especially for non-IPs CSOs.

2.5 Alternative Explanations

The adoption of a particular frame in climate advocacy has been approached from various theoretical angles. We argue that our theoretical framework provides a better understanding of CSO populations that are often overlooked in the extant literature on climate advocacy. In particular, our approach gives us a sense of how small and specialist CSOs can lead framing efforts and shape our understanding of climate change despite their relatively disadvantaged political positions at the UNFCCC.

One dominant approach highlights the positions of CSOs in their advocacy networks. Scholars argue that CSOs with strong connections to peer organizations serve as "gatekeepers," who have the ultimate power to decide advocacy agendas (Bob, 2005; Carpenter, 2007). Similarly, scholars point out that small CSOs free-ride on the advocacy resources that prominent CSOs provide (Murdie, 2014). In the context of climate advocacy, Allan (2020) argues that "lead NGOs" connect different clusters of CSOs to broaden issue-based coalitions. Overall, existing research suggests that CSOs positioned at the center of an advocacy network can control how climate change should be framed because other CSOs would bandwagon on their advocacy agendas.

We argue that connecting to peer organizations may be *one way* to popularize a particular frame at the UNFCCC. For example, as we detail in empirical sections, well-networked CSOs can impose the idea of what should be appropriate forms of advocacy at the UNFCCC, which small CSOs must accept willingly or unwillingly in order to participate in the UNFCCC process. However, interorganizational networks are among many kinds of resources that CSOs can leverage in order to advance their advocacy positions. Recent research on CSOs also provides empirical evidence that smaller CSOs can set a new agenda without connecting to "gatekeepers" (Pallas and Nguyen, 2018; Shibaike, 2022). Others also highlight the limits of resource-rich NGOs in taking leadership in advocacy work (Balboa, 2018; Stroup and Wong, 2017). In the context of climate advocacy, while Allan (2020) argues that the presence of "lead NGOs" allowed marginalized groups to participate in the UNFCCC, Hadden (2015) shows that prominent CSOs like the Friends of the Earth did not act as brokers between climate networks.

The gatekeeper thesis may explain some cases but not others, suggesting the need for further theorization of CSOs. Except for a few recent studies, research has drawn heavily on the experiences of prominent CSOs and neglected "other" CSOs. In other words, we do not know why some ideas or frames proliferate in a given context when gatekeepers refrain from exercising such power. For example, we did not find any gatekeeping activities by IPs' organizations, but indigenous-climate frames are increasingly popularized at the UNFCCC.

Instead of gatekeepers, we need to focus on the strategic concerns surrounding all CSOs in the advocacy space.

Another strand of research focuses on the power of framing itself. A successful frame fits well with the existing social and political contexts such that audiences accept it as a legitimate concern (Bernstein, 2001; Price, 1998). Certainly, CSOs exercise reflexive interpretations to produce frames that they hope to be most effective. For example, Allan and Hadden (2017) argue that justice framing broadened the coalition of CSOs at the UNFCCC and ultimately raised awareness of "loss and damage" concerns. Focusing on ideational power, scholars of climate governmentality argue that frames are produced by the logic of eco-modernization and the supremacy of science primarily associated with industrialized nation-states (Bäckstrand and Lövbrand, 2006; Dowling, 2010).

The issue with this approach is that we can provide an account for advocacy success, only after the frame or issue becomes a major concern. For example, advocates for forest conservation might use a scientific frame to justify a particular deforestation technique, but it can also feature IPs as defenders of forest. Since both science and IPs are relatively prominent themes at the UNFCCC, the fit with existing social structures cannot explain organizational choices to adopt one frame over the other. Our approach points to the density of organizations as well as the disciplinary process to explain why CSOs adopt a particular frame among a range of possible options.

Finally, the principal-agent theory provides an illustration of representation problems, where an agent, who is elected to and claims to represent a principal (the electorates), does not provide sufficient and effective representation for the principal (Olson, 1971). Accountability problems of CSOs have been frequently discussed in the context of relief and developmental aid allocations (Cooley and Ron, 2002; Gent et al., 2015). In climate governance, the interest of people groups like women and IPs is represented in the UNFCCC through CSOs. Although it is possible to conceptualize the relationship between people groups and CSOs as principals and agents, we do not have a good metric to evaluate the "slippage" of agents, which is often the central focus of principal-agent research. One may argue that conservation CSOs using IPs frames in their advocacy deviate from their original missions, but such a normative evaluation falls outside the scope of our study.

Moreover, the principal-agent framework requires a clear understanding of interests in both principals and agents. As we detail in Section 4, however, what benefits the interest of people groups is not always clear in the context of climate governance. For example, IPs are divided whether REDD+ initiatives contribute to their welfare and independence, and the conditions for gender equality do not have a consensus among different actors. As scholars of political representation argue, the interest of people groups and CSOs claiming to represent their interest are co-constitutive (Disch, 2012; Saward, 2006), which cannot be accounted by the principal-agent theory.

To overcome these shortcomings discussed previously, our theory focuses on both organizations and frames without reducing one to the other. More concretely, the framework of organizational ecology allows us to think about what types of CSOs go into the ecosystem of the UNFCCC advocacy space. It also lets us think about the frames as strategies to extract resources from the audiences and observe the distinct advocacy space for each CSO population. In other words, our approach bridges the gap between organization-based and discourse-based analyses.

2.6 Conclusion

This section provided a theoretical framework for the empirical analyses in the following sections. We have argued that CSOs use various frames to extract social and material resources at the UNFCCC. In particular, CSOs specialize in a narrow niche in order to maximize the chance of survival and organizational growth. We have provided two ways in which specialist CSOs develop their frames: self-representation and surrogate-representation. These different patterns of niche occupation can have far-reaching consequences for the future of climate advocacy at the UNFCCC.

Our theory contributes to the broader discussion of the nature of CSOs in transnational advocacy. The notion that such actors are normatively principled actors has been challenged by the empirical findings that they are also political actors maximizing self-interest in a resource-scarce world (Bob, 2005; Cooley and Ron, 2002; Keck and Sikkink, 1998). Our study contributes to this long-standing debate by arguing that CSOs are maintaining careful balance between principled commitments and resource security for organizational survival. In the following sections, we will demonstrate that specialist CSOs – as opposed to resource-rich, generalist CSOs – occupy distinct discursive space through self- and surrogate-representations, which ultimately shape dominant ideas about climate change at the UNFCCC.

3 The Analysis of CSO Framing Efforts on Social Media

This section examines how CSOs participate in and shape climate change discourse at COP21 and demonstrates the important role of specialist CSOs. By specialist CSOs, we mean CSOs who focus on specific issues and people groups. Generalist CSOs have multiple issue interests related to climate change.

We draw on the theoretical framework of organizational ecology and demonstrate how CSOs occupy climate niches at the UNFCCC and how such organizational dynamics ultimately shapes our understanding of climate change. In particular, we show empirically how self-representation and surrogate-representation occur in the areas of women's advocacy and IPs' advocacy. We use Twitter (now X) data from registered non-Party observers at COP21, a pivotal moment in global climate governance with the reaching of the Paris Agreement.

We aim to achieve three things in this section. First, we empirically and visually illustrate the multidimensional nature of framing efforts. Existing research has often reduced the diversity of frames in climate change advocacy to the sectors that CSOs belong to (business, gender and women, youth, etc.) based on the UNFCCC's constituencies and thematic groups (Allan, 2020; Cabré, 2011). However, simply tallying the number of CSOs does not tell us how and why various CSOs adopt particular frames in practice. By leveraging Twitter data and computational methods, we present the first systematic effort to capture variations in both the organizational attributes of CSOs *and* their framing efforts. In so doing, we locate the relative positions of CSOs in a discursive space at COP21.

Second, we examine the relationship between organizational attributes and different framing choices. To date, research on CSOs has suggested that a handful of actors serve as gatekeepers in agenda-setting processes while small CSOs jump onto the bandwagon of popular causes (Bob, 2005; Carpenter, 2007). We challenge this notion by adopting the framework of organizational ecology and explain how organizational characteristics may or may not affect different framing choices of CSOs.

Finally, we highlight the utility of leveraging online data as a study of transnational activism. While the UNFCCC has opened the door to non-Party observers, most of them still do not impact negotiation outcomes directly. After all, the UNFCCC is an inter*governmental* institution, where national delegates meet and negotiate climate-related issues. To be sure, a small portion of CSOs can join the national delegations to have their input during diplomatic negotiations (Betsill and Corell, 2008). However, the vast majority of them are kept outside of official decision-making processes. Online space allows us to observe the work of CSOs without being dependent on gatekeepers' decisions.

We focus our data collection efforts on Twitter, which is (or, was before Elon Musk's acquisition of the company) among the most widely used social media platforms for political communication. As we are interested in framing choices, the brevity of tweets (140 words at the time of COP21) helps us identify one frame for each tweet. Merry (2013) finds that environmental CSOs generated sustained attention on environmental issues on Twitter than traditional media. In 2015, the Pew Research Center found that roughly 60% of US adult respondents used Twitter as a way to learn real-time news events (Shearer et al., 2015). Research in the United States also suggests that active Twitter users are politically engaging and can shape policy discourse (Hemphill and Roback, 2014). In short, Twitter provides a useful forum to observe how CSOs engage in climate advocacy.

3.1 The Paris COP

3.1.1 Data Collection and Processing

We focused our data collection efforts on COP21 in Paris, a milestone event in global climate governance. We posit that CSOs were highly active in disseminating their messages during COP21, for which there was a strong expectation that the meeting would produce a new international agreement on climate change. In fact, the conference witnessed the highest number of non-state observers at the time, with a total of 1,109 participants. This record turnout of participants was only exceeded during COP26 in 2021.

To analyze how different CSOs used different frames for climate change, we assembled two types of data for *all* registered CSO participants for COP21. First, we collected all English-language tweets from the registered observers, including retweets, during the entire 14-day period of COP21 (November 30–December 12, 2015). We used the hashtag #COP21 as a keyword to focus on CSOs' intentional efforts to shape climate change discourse during COP21. CSOs are also much more likely to use hashtags than ordinary users (Boyd, Golder, and Lotan, 2010). Although some CSOs do not have Twitter accounts or do not actively engage in online discussions, such negative data (i.e. the dog that didn't bark) are also valuable in evaluating CSOs' representation in climate change discourse. We collected a total of 20,851 tweets.

We coded the tweets based on the climate change frames represented by each tweet. Since there has been a proliferation of frames in climate change discourse in recent years (Jinnah, 2011), different studies use different numbers of frames in their analyses (Allan and Hadden, 2017; Hjerpe and Buhr, 2014; Hopke and Hestres, 2018). As we are interested in the different mechanisms under which *gender* and *indigenous* frames are established, these two categories are included in our coding scheme. We also included well-established frames identified in the literature: *science*, *energy*, and *economy*. Finally, we added two categories specific to COP21: *invite* (event invitation) and *treaty*.

To code a large number of tweets, we adopted the semi-supervised machine-learning (ML) model called the seeded Latent Dirichlet allocation (LDA). The LDA is a method to classify documents, such as tweets and newspapers, into any number of groups (or clusters) based on the probability distribution of each word in the text (Bagozzi, 2015; Blei, Ng, and Jordan, 2003). For example, if relatively unique words, such as "indigenous" and "native," appear across a subset of documents, these documents will be clustered together.[9] In other words, words like "climate" will not be a useful identifier for clusters because a large majority of tweets include "climate" regardless of the frame they use.

A major problem of unsupervised-ML (non-seeded LDA) is that the interpretation of the results is necessarily ad hoc. Researchers can specify the number of clusters, but they must ensure that each of the clusters mean anything useful *after* the model outputs results. The seeded LDA has an advantage over non-seeded LDA models in that it allows us to specify keywords – seed words – for document classification based on our prior knowledge. Here, the model uses the co-occurrence of seed and non-seed words for document classification (Ramesh et al., 2014; Watanabe and Zhou, 2022). The seeded LDA finds the words that are likely to co-occur with seed words and group together the documents that use them. The seeded LDA also offers an advantage over supervised ML models, which are heavily influenced by our selection and coding of "training" sets (Hastie et al., 2009; Yu, Kaufmann, and Diermeier, 2008). In short, seeded LDA attempts to take a middle ground between supervised and unsupervised ML models.

We chose three to four words as the seed words of each frame in our model.[10] The model then sorted each tweet into one of the eight categories based on the seed and co-occurring words. Note that, in addition to the seven categories discussed previously, the model added a residual category, *other*, to handle ambiguous cases. Table 1 shows the top ten words most strongly associated with each frame and a total number of tweets sorted in each frame category.

Second, we collected organization-level data by web-searching every CSO on the list of admitted non-state observers (1,109 organizations) at the COP21. We collected the mission statement, the geographic scope of operation, and headquarter locations, from each CSO's official website. If a CSO has no online presence and thus no information about the organization, we dropped them

[9] To be sure, whether they are actually clustered together depends on the number of clusters we want to find in the text.

[10] science = target*, carbon, emission*, redd*; energy = fossil, fuel, oil, nuclear*; economy = economy, market*, business*; gender = woman, women, gender; indigenous = indigenous, native; invite = event, join, watch, follow; treaty = paris, deal, agreement, talk*

Table 1 Top ten words most strongly associated with each frame and a total number of tweets in each category. Seed words are indicated by gray shades

	Science	Energy	Economy	Gender	Indigenous	Invite	Treaty	Other
1	carbon	fossil	business	women	indigenous	event	paris	cities
2	emissions	nuclear	economy	gender	climatechange	join	agreement	action
3	target	nuclear4climate	businesses	world	change	watch	deal	world
4	redd	fuel	markets	people	rights	follow	talks	leaders
5	targets	oil	market	action	can	today	talk	statesandregions
6	emission	countries	marketsmatter	leaders	forests	live	talking	ban
7	finance	need	energy	now	birdstellus	us	text	lpaa
8	can	us	clean	future	need	day	new	local
9	adaptation	support	climatetv	climatejustice	health	now	draft	global
10	role	1o5c	renewable	justice	must	side	parisagreement	ki-moon
#	2,562	2,503	2,590	2,900	2,795	2,934	3,101	2,618

from the analysis. We collected organization-level data for a total of 834 CSOs. Based on the mission statements, we identified the sectors to which each CSO belonged in the context of climate governance. We used Cabré's (2011) CSO sectors at the UNFCCC, such as "Business & industry" and "Rights & justice," as a starting point. We then added a few other categories for COP21 to update Cabré's categorization based on COP15 (2009). The added sectors include "Conservation," "Finance & market mechanism," "Health," and "Pollution & waste."

In addition, to operationalize the concept of specialization in the context of environmental and climate CSOs, we took into account the scope of each CSO's mission and categorized them into three groups. Each CSO was categorized into one of the groups. We treated organizations such as universities, businesses, and churches as separate categories. Table 2 summarizes the statistics of CSO categorization. A full list of CSO sectors and summary can be found in Appendix A.

1. **Multi-issue CSOs**: *Generalist* CSOs that cover a broad range of issues belong to the climate change and environmental sectors. For example, Greenpeace International is coded as a multi-issue CSO, as its mission statement encompasses a wide range of issues, including forest, ocean, air and water quality, and food security, among many others.
2. **Issue-specific CSOs**: *Specialist* CSOs with a specific environmental focus, such as food, agriculture, energy, forest, health, transport, water, and ocean, belong to this category. Kenya-based African Forest Forum is an example of issue-specific CSOs. In this case, the mission of the CSO focuses specifically on Africa's forest and tree resources.
3. **People-specific CSOs**: *Specialist* CSOs with a focus on the welfare of a specific people group, such as women, youth, and indigenous peoples, belong to this category. For example, GenderCC, a women's CSO in Germany, is coded as a people-specific CSO. Its endeavor is exclusively in gender equality and women's rights in battling climate change.

Regarding geographic information of CSOs, we focused on two kinds of organizational data. First, the operating scope indicates the primary areas of operation for each CSO. The scope has three levels: whether its activities take place locally and nationally, regionally (such as within the European Union or in Southeast Asia), or globally. Second, we coded whether each CSO's headquarter is located in the global North or South. Table 3 summarizes the geographic information of CSOs.

Table 2 Categorization of CSOs generalists and specialists according to sectors

	Number of CSOs	Percentage (%)	Sectors included
Multi-issue CSOs	142	17	Climate change, Environment
Issue-specific CSOs	223	26.7	Built environment, Conservation, Energy, Finance and market mechanism, Forest, Food and agriculture, Health, Legal practice, Pollution and waste, Rights and justice, Sustainable development, Transport, and Water and oceans
People-specific CSOs	70	8.4	Indigenous peoples, Women, Youth and children
University	140	16.8	University
Business and industry	67	8	Business and industry
Others	182	21.8	Education and capacity building, Development, Religious and spiritual, Science and engineering, Think tank, Others and unknown

Table 3 Headquarters and operation scopes of CSOs

		Number of CSOs	Percentage (%)
Headquarter	North	689	82.6
	South	143	17.2
	Unknown	2	0.2
Operating scope	Global	481	57.7
	Regional	83	10
	National	223	26.7
	Unknown	47	5.6

3.2 Analysis

3.2.1 Discursive Map of CSOs

Our coding of tweets allows us to empirically observe the framing choices of CSOs without reducing them to CSOs' sector-based attributes. For example, Greenpeace International is a generalist, multi-issue CSO. We can measure what are bundled in this "multi-issue" package by analyzing the composition of its tweets: 9 tweets with scientific frames, 54 with gender, 26 with indigenous, and so on. We posit that this variation indicates how Greenpeace International distributed its advocacy efforts between different climate change frames during COP21.

CSOs tweeted with various frequencies during the COP. Among 834 CSOs, 458 CSOs tweeted at least once, ranging from a single tweet to 1,540 tweets. Figure 1 shows the distribution of the total number of tweets by CSOs. As there is a variation in CSOs' tweet frequency, we standardized framing efforts with a total number of tweets for each CSO. In short, *framing variables* measure the *ratio* of tweets adopting each of the eight frames introduced earlier (Appendix B). In this analysis, we removed CSOs with 14 or fewer tweets (i.e. one tweet per day on average), so extreme values are generated by framing choices rather than the small number of total tweets.[11] A total of 213 CSOs are analyzed.

We used these eight framing variables (corresponding to Table 1) to represent the positions of CSOs in the discursive space of climate activism during COP21. More concretely, we treat each framing variable as a dimension in the discursive space (with a total of eight dimensions). Although impossible

[11] For example, if a CSO tweets five times during the COP, four of which were about science, this CSO would receive a high value, 80%, for scientific framing.

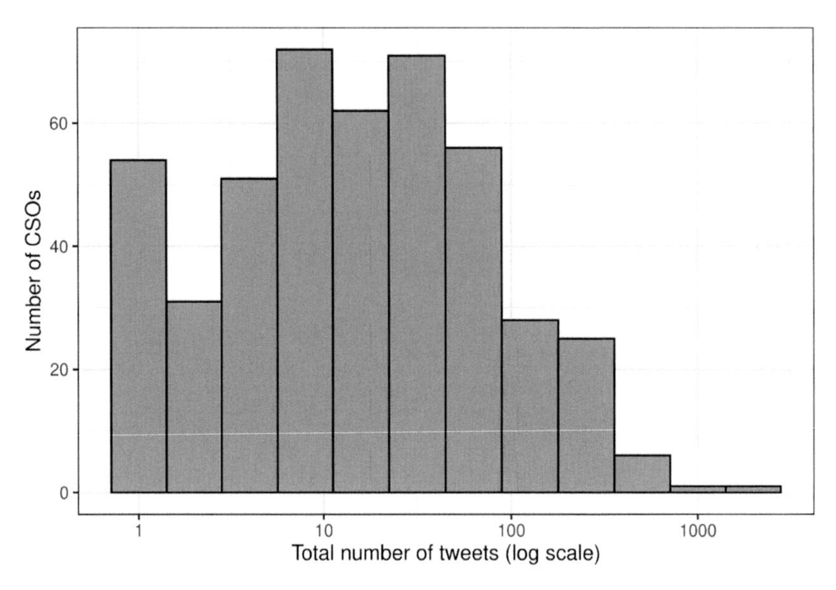

Figure 1 Histogram of total number of tweets by CSOs

to visualize (as there are eight dimensions), each CSO's position is determined by the eight framing variables. Doing so allows us to compare the similarity of CSOs in terms of actual tweets, completely independent of organizational attributes. To provide an intuitive visualization, we plotted each CSO on a two-dimensional space with multidimensional scaling (MDS). MDS is a method to reduce any number of dimensions to a useful number while preserving similarity among units as much as possible (Kruskal and Wish, 1978). In our case, MDS reduced eight to two dimensions so that we can visualize the relative positions of CSOs on the plane. In Figure 2, CSOs close to each other share similar compositions of framing choices. By contrast, CSOs far apart from each other are dissimilar in terms of framing choices. At the center of this discursive map are "average" CSOs, such as Sierra Club, Practical Action, and the United Church of Canada, leaning more toward pragmatic approaches than radical actions in their climate advocacy. By contrast, the further a CSO is from the center, the CSO is more "extreme" in terms of focusing on a particular framing strategy.

To visualize the clusters of CSOs (i.e. groups that share similar frames), we conducted a k-means clustering analysis on the eight framing variables. The k-means cluster analysis is a method to find clusters (groups) of units by minimizing the distance to units from the center of each cluster (Bode et al., 2015). Here, we assume that there are five important clusters because, among the

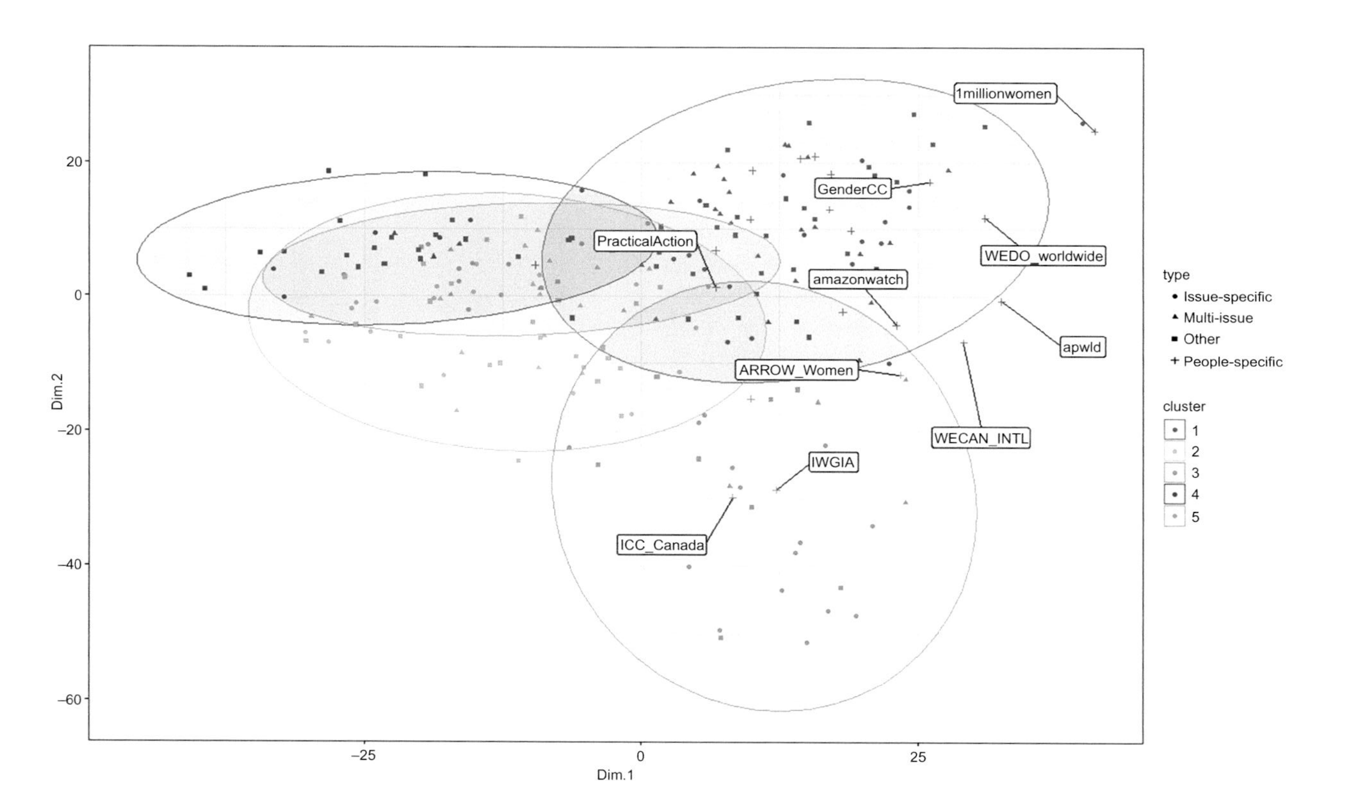

Figure 2 Discursive map of CSOs with five clusters indicated by color circles. Labels indicate CSOs (Twitter handles) that belong to women's and IPs sectors. Node shapes represent generalist and specialist CSO categories. The distance between any given CSOs indicates the similarity of framing choices during the COP21.

eight framing variables, five are substantive topics (science, energy, economy, gender, indigenous). Doing so allows us to see which organizations belong to which cluster (i.e. share similar framing choices) on the discursive map. This grouping is important and novel because it does not rely on organizational attributes or the UNFCCC constituency that existing research frequently uses. The clusters are visualized in color in Figure 2.

We find three major groupings in this discursive space. First, the orange circle (Cluster 5) consists of CSOs that focus mainly on gender frames. We find that specialist CSOs representing the interest of women, such as GenderCC, WEDO (Women's Environment & Development Organization), 1 Million Women, are "pulling" this circle toward the upper right. As gender issues constitute a well-established niche populated with various CSOs, these women's CSOs must distinguish themselves by spearheading their own narrative in the context of climate change. We also find a concentration of people-specific CSOs, suggesting that they may find easier ways to connect their group identities, such as youth and students, with women and gender.

Second, the blue circle (Cluster 2) consists of CSOs tweeting mainly with indigenous frames. Unlike the gender cluster, IPs organizations, such as ICC Canada and IWGIA (International Work Group for Indigenous Affairs), are located in the center of the circle rather than on the fringe. CSOs at the bottom of this circle are specialist CSOs focusing on specific issues, such as forest and health – issues that are connected to IPs. This suggests that a different set of specialist CSOs, the issue-specific ones, rather than IPs' organizations, are pushing the narrative of IPs. Moreover, this IPs' cluster occupies a distinct area in this discursive space, suggesting that their framing profiles are unique and potentially making it difficult to collaborate with the rest of the CSO participants in COP21.

Finally, three clusters (Clusters 1, 3, and 4) overlap significantly on the left side of the quadrant, suggesting that CSOs in this discursive space share similar framing choices. These are a mix of CSOs that focus on traditional frames, such as science, energy, and economy. Toward the left, we find research institutes, such as Edison Electric Institute and IGSD (Institute for Governance & Sustainable Development), that focus heavily on scientific frames.

We also measured the distance of each CSO from the mainstream climate change discourse at COP21. The mainstream climate discourse is defined here as the *average* of each framing variable. Drawing on the strategy of measuring ideological distances developed by Baldassarri and Bearman (2007), we interpret this distance as a measure of *extremity* for CSOs' framing choices. That is, the further a CSO deviates from the average framing choices, the more extreme is its framing strategy relative to the rest of CSO participants at COP21. On the discursive map, CSOs at the fringe tend to have high extremity. Formally, the distance (D) of each CSO i is computed as follows:

$$D_i = \sqrt{\sum_{n=1}^{j} (p_{i_j} - q_j)^2} \tag{3.1}$$

where p is each CSO's (i) framing variable (the ratio of each frame adoption) and q is the global average of each framing variable j.

We used an Ordinary Least Squares (OLS) regression model to evaluate the relationship between the framing extremity (D) of each CSO and its organizational attributes. This analysis shows us what kind of organizations are "pulling" climate discourse in their own directions rather than trying to appeal to the largest possible number of audiences. We found that specialist CSOs focusing on particular issues were more likely to be extreme in their framing choices relative to generalist CSOs (Appendix C). The result suggests that generalist CSOs occupy the middle ground in climate change discourse at COP21, using largest-common-denominator strategies in framing efforts. Interestingly, specialist CSOs focusing on people groups are positively associated with the extremity measure but statistically insignificant ($p = 0.08$). The result could mean two things. First, frames that people-specific CSOs adopt are considered mainstream discourse at the UNFCCC. Given the strong presence of people-specific CSOs in the gender cluster (Figure 2), this could be another indication of the mainstreaming of gender discourse in global governance (Hafner-Burton and Pollack, 2002). Second, it is also possible that people-specific CSOs are taking more pragmatic approaches than issue-specific CSOs to advance their agendas. The finding here highlights different ways in which specialist CSOs are represented in the discursive space of COP21 and shows how specialization alone cannot predict the extremity of specialist CSOs.

3.2.2 Who Contributes to Climate Change Discourse?

Having established that framing choices are not necessarily reducible to organizational attributes, we take a close look at the relationship between organizational attributes and framing choices. First, we examine what kind of CSOs were more likely to engage in online activism during COP21. We used the total number of tweets during COP21 for each CSO to measure online engagement. We applied negative binomial models (Gardner, Mulvey, and Shaw, 1995) to fit our event-count dependent variable (i.e. the number of tweets). We first constructed a baseline model, predicting the number of tweets with a categorical variable for the CSO type (multi-issue, issue-specific, people-specific, or other CSOs), operating scope (national, regional, or global), and location of CSO

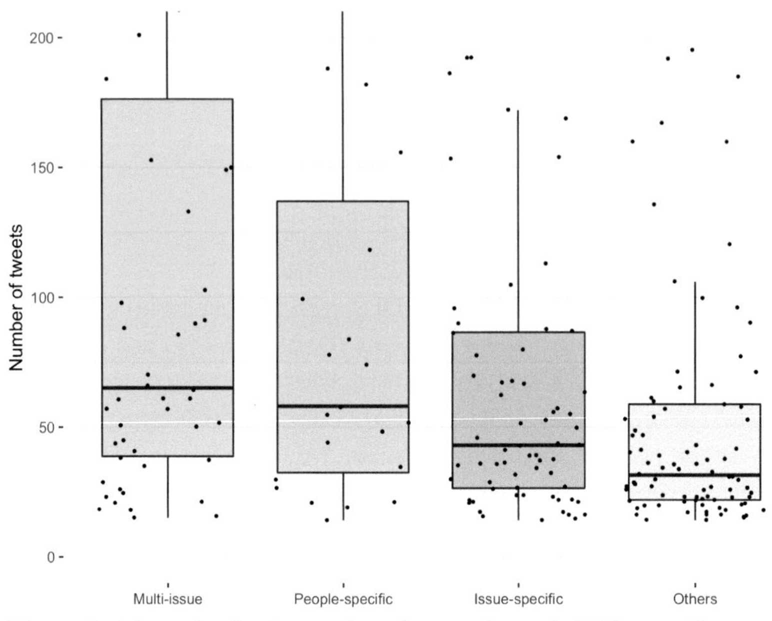

Figure 3 A box plot for the number of tweets by each CSO type. The sample includes CSOs with 14 or more tweets, excluding outliers for visualization.

headquarter (South or North) (Appendix D). We then added two specifications (Appendix E) where Model 1 replaces the CSO type variable with a binary variable for each of the CSO types, and Model 2 includes specific sectors.

The baseline model shows that generalist CSOs that focus on multiple issues are the most active actors in climate change tweets. Using multi-issue CSOs as a reference category, we find that all other categories of CSOs show a negative coefficient. The result indicates that specialist CSOs are less active in tweeting during COP21 than the generalist CSOs. To visualize this effect, Figure 3 shows the number of tweets from each CSO type. CSOs whose work covers an expansive geographic region and those from the global North are more likely to be active tweeters. Model 1 is consistent with the baseline model, showing that multi-issue CSOs tweet most actively, and Northern CSOs are more likely than Southern CSOs to be active on Twitter. According to the baseline model (Figure 3) and the marginal effect analysis (Figure 4, Panel a), generalist CSOs are likely to tweet more than the other types of CSOs.

The findings here offer two important insights. First, generalist CSOs have a greater capacity than specialist CSOs to appeal to a broader set of audiences, as they have can pick and choose from a wide variety of climate change frames. Among the top ten tweeters during COP21, seven are multi-issue generalist CSOs working in climate and environmental sectors, and all are based in either

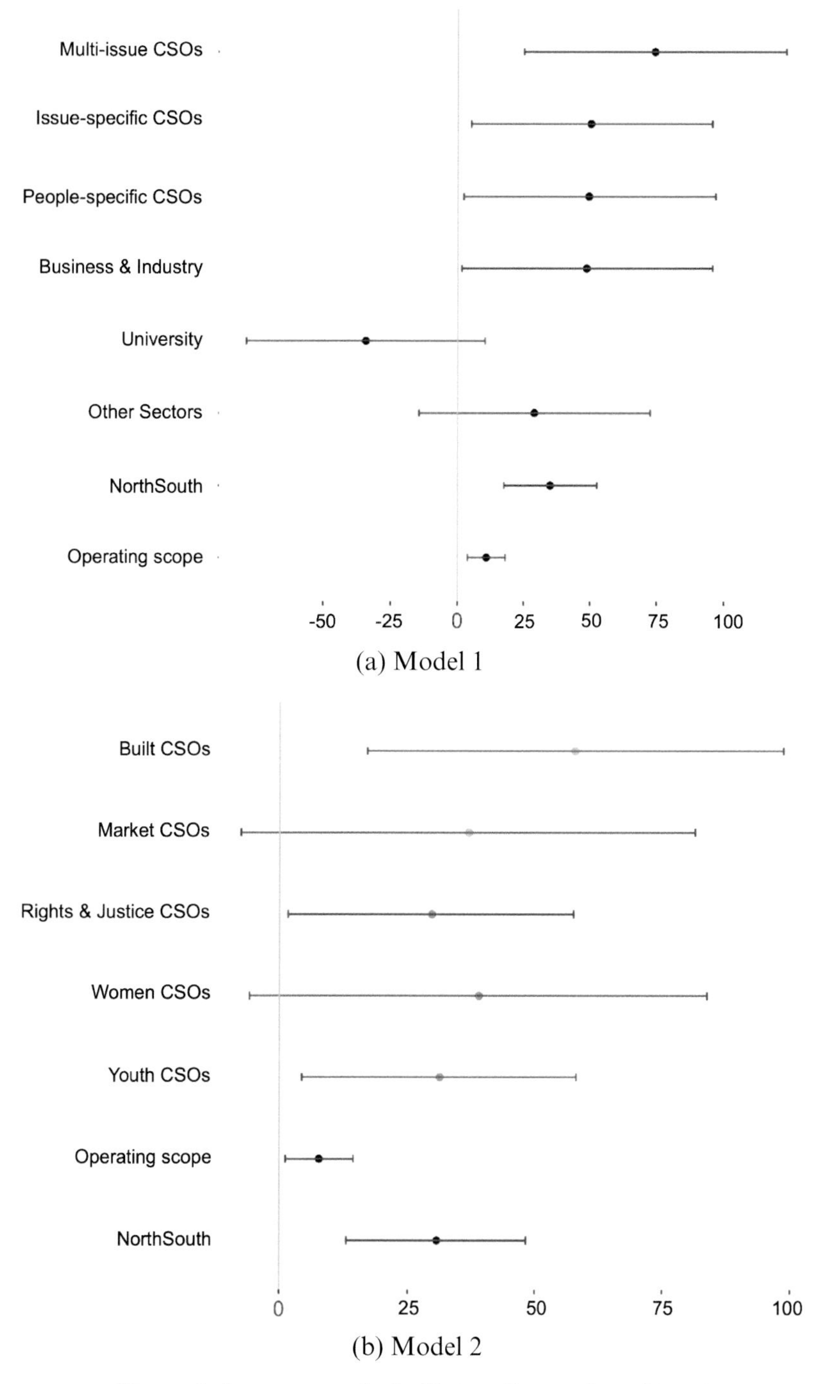

(a) Model 1

(b) Model 2

Figure 4 Average marginal effect on the number of tweets

Europe or the United States. Backed by relatively abundant resources, these generalist CSOs can create what can be seen as "mainstream" discourse in climate change activism.

Second, the findings also illustrate the lack of representation by Southern CSOs. The UNFCCC COPs are known to be more accessible venue than many other multilateral processes, where the participation of non-state actors is restricted. While CSOs may find it relatively easy to register as attendees, not everyone contributes to the construction of climate change discourse. Generalist CSOs that are based in the global North and work in a variety of geographic areas are far more likely to contribute to climate change discourse than their global Southern peers. The North-South asymmetry in global governance has been criticized by many (Barnett and Walker, 2015), but we also find this asymmetry in the social media space.

However, we know from the previous analysis (Figure 2) that climate change discourse is fragmented and is not dominated by large generalist CSOs. Specialist CSOs, especially those that engage with gender frames, have a strong presence in the overall discourse of climate change. Although specialist CSOs tweet less frequently than generalist CSOs, this does not mean that they are absent or even marginalized. Therefore, Model 2 unpacks the two specialist categories (issue-specific and people-specific CSOs) and uses CSO sector attributes as estimators.

Model 2 shows that five sectors of specialist CSOs are particularly active on Twitter during COP21: the built environment community, climate financing and market mechanisms, environmental rights and justice, women, and youth. The average marginal effect of each sector is plotted in Panel b of Figure 4. Most CSOs working in the first two sectors – built environment such as urban management and planning and market mechanisms such as carbon trade – are from the global North. While these CSOs are specialists in certain issues within climate change, their sectors have so far been dominated by Northern expertise. Their activeness supports the previous finding that Northern groups enjoy higher online engagement.

However, the other three sectors – rights and justice, women, and youth groups – include many Southern organizations and groups, and, more importantly, represent the interest of disadvantaged social groups. These CSOs are specialized in social issues related to justice and equity, the frames that recently became popularized at UNFCCC. Their clear presence on Twitter signals the rise of CSOs that do not derive their expertise from scientific or technological knowledge, but from the ability to access and represent the people who are disproportionately affected by climate change. While large generalist CSOs do enjoy privileged positions, being specialists is also a source of power in engaging with climate change discourse. Our analysis thus joins the recent effort to re-imagine a dichotomous portrait of the powerful gatekeepers versus their followers among CSOs (Shibaike, 2022; Zhao, 2023), providing a nuanced picture of an ecological space of climate framing where different groups can thrive at the same time.

3.2.3 Stories We Tell: Gender Advocacy

The previous section established that specialist CSOs are important contributors to climate change discourse, but it did not tell us how they frame climate change. As women and gender advocacy continues to gain attention at the UNFCCC, we investigated what kind of groups developed this particular frame during COP21. We used the gender framing variable (one of the framing variables in Appendix B), which is the ratio of gender frames each CSO chose out of all tweets. In other words, this measures the extent of efforts that each CSO made in constructing gender framing during COP21. As the percentage data is bounded between 0 and 1, we used binomial generalized linear models to predict the outcomes (Appendix F).

As the frame is primarily around women and gender topics, we first tested whether women's CSOs are active in constructing this frame. Model 1 from Appendix F shows that women's CSOs are highly active in choosing gender frames. Although only 18 CSOs belonged to the women's sector based on an explicit commitment to the cause in their mission statements, almost all of them dedicated their COP21 tweets to advocating women's rights and gender equality. Eight of the 18 women's CSOs have at least one-third of their tweets focusing on women and gender issues. Those women's CSOs together formed a community that is highly vocal about their activities and goals at COP21, representing the interest of women from both the global North and South. Although the North-South divide is a persistent problem among most sectors, we find little evidence of that among women's CSOs.

We also explored whether CSOs from other sectors contribute to women and gender frames compared to women's CSOs. Model 2 expanded Model 1 to include generalist and specialist categories; people-specific CSOs are not included, as women's CSOs exclusively belong to this category. As the marginal effects in Figure 5 show, CSOs from no other types stand to be noteworthy co-contributors of gender frames. In particular, although generalist CSOs could have taken on gender issues as a major part of their multi-issue agenda, we found no evidence that they concentrated their advocacy resources on this frame. Generalists avoid overcommitting to gender framing despite being active contributors of climate change discourse. Instead of generalists, the prominent voice comes from women themselves.

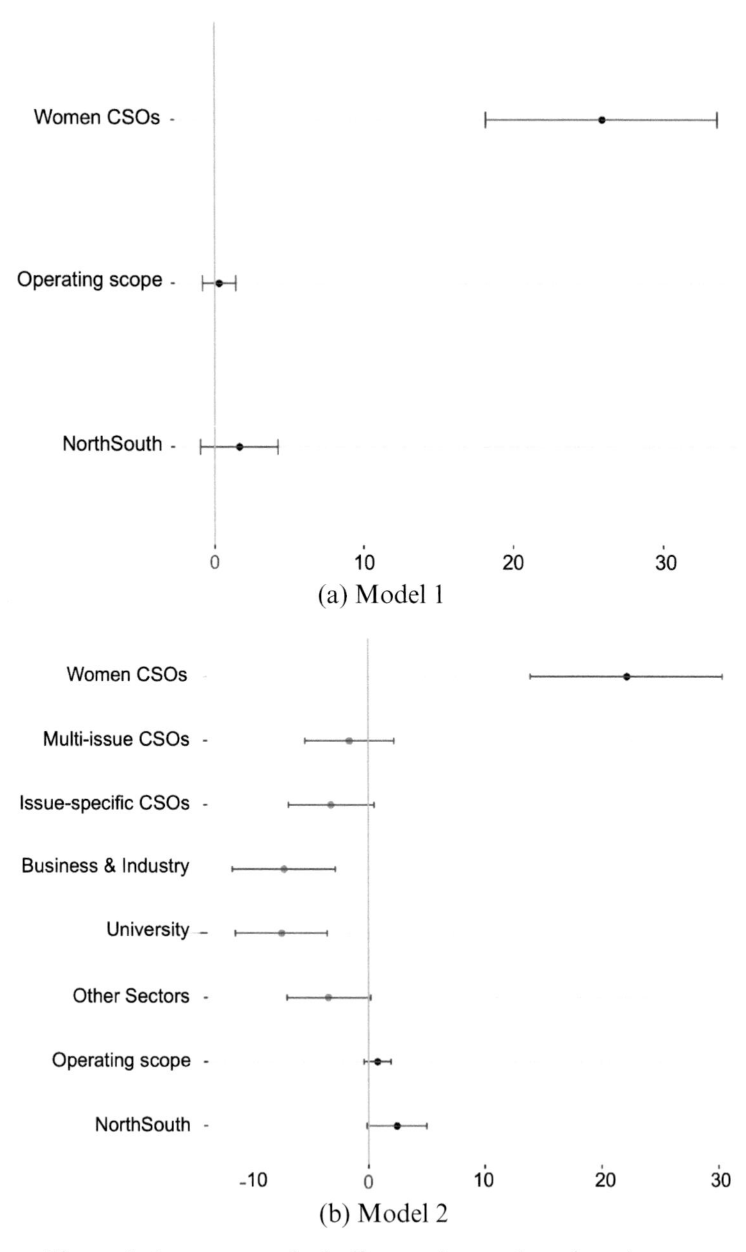

Figure 5 Average marginal effect on the number of gender tweets

The findings have three implications. First, women's CSOs not only work inside the UNFCCC venues to have themselves heard, but also use social media to amplify their voices. Huairou Commission, a grasssroots women's movement, had all their tweets tightly surrounding women and gender issues.

1 Million Women from Australia adopted a gender frame for 75% of their tweets at COP 21. The tweets from these CSOs and other women's SCOs did not shy away from their agendas. They demanded representation: "If you are not at the table you are often on the menu. We need more #women at the #cop21 table. #ClimateJustice" (1 Million Women). They also called for mainstreaming gender into the formal agreement: "#COP21: Gender is cross-cutting & must be included in Purpose Section in the agreement! #WhatIsThePurpose #WEDOecho" and "You cannot trade off half the population on women and gender issues disappearing at #COP21" (WEDO). They portrayed women as agents of change: "Women bring solutions both at COP and on the ground" (WEDO). A clear message insisting on gender balance and women's rights is found in many of their tweets.

Second, a clear division of labor is emerging between CSOs constructing and shaping women and gender frames and those adopting them as a part of, but not necessarily the center of, their advocacy agendas. Women's CSOs are committed to shaping their own stories. They are the main drivers of the agenda. Other CSOs do not exhibit a consistent effort to adopt the gender perspective and be an advocate for women's rights and gender equality in their narratives. This supports our idea about self-representation, in which a populated niche pushes specialist CSOs to develop genuine ownership of an issue – gender in this case.

Finally, the strong presence of networks among women's CSOs may have contributed to the formation of similar framing strategies among them. As we discussed in Section 2 and will detail in Section 4, the WGC had a disciplinary process in developing a unified voice among CSOs. This mechanism is further strengthened by network organizations like GenderCC, in which member CSOs share similar strategies.

3.2.4 Stories They Tell: IPs Advocacy

We investigated another group of people-specific CSOs, IPs' organizations, to understand their framing efforts. Just like previously, we used the indigenous framing variable (Appendix B) as the outcome variable and examined how different types of CSOs contributed to the development of indigenous frames in climate change discourse. The results from statistical models are shown in Appendix G.

We constructed Model 1 to test whether IPs themselves are active contributors to indigenous frames. We find no evidence that IPs' organizations spent significant portions of their tweets about their own stories. Moreover, the substantive effect of the IPs sector on indigenous framing is negative. These results indicate that IPs' organizations are no more likely to use indigenous frames than CSOs from other sectors. Forty CSOs have at least half of their COP21 tweets devoted to indigenous framing. Only one IPs' organization, Asia Indigenous Peoples Pact Foundation, was found among them.

Other IPs organizations were also active on Twitter, but they did not engage with IPs' advocacy with an effective communication strategy that other SCOs did (as we see next). For example, the ICC Canada tweeted an update on their activity, "ICC meets with US Secretary of the Interior, Sally Jewell #COP21 #climatechange." Others were invitees to the events that they or their allies hosted: "Join #IndigenousPeoples Pavilion at COP21 Grand Opening Ceremony 1 Dec 14.15" (IWGIA) and "Happy International #Youth and Future Generations Day" (ICC Canada). While IPs' organizations did tweet about themselves, the ratio of such tweets was relatively low.

An optimistic interpretation about this finding is the mainstreaming of indigenous discourse: many non-indigenous CSOs not only chose indigenous framing for their climate advocacy but also focused on it during COP21, so IPs organizations were not alone in talking about their climate concerns. However, there is also a possibility of tokenism in which stories about IPs were appropriated by other CSOs to create emotional appeals.

To look more closely at organizational attributes in relation to IPs' advocacy, we expanded our model to include CSO types (multi-issue, issue-specific, and other CSOs) in Model 2 and all sectors of the registered CSOs in Model 3. Model 2 shows that issue-specific CSOs, instead of IPs organizations, are helping to build an indigenous-climate narrative. Model 3 further corroborates the finding that some of the issue-specific specialist CSOs are dedicated to constructing indigenous framing for climate change. Panel c of Figure 6 shows the sectors of CSOs actively using indigenous frames: conservation, forest, environmental health, and rights and justice.

Global Witness, the UK-based organization that seeks justice for environmental and climate defenders, is an example of such issue-specific CSOs. Its mission statement emphasizes the people who are not necessarily IPs but ill-treated or murdered by governments and corporations because of their efforts to protect the environment and fight against climate-wracking industries.[12] Global Witness is committed to forest and biodiversity conservation and justice for those disproportionately affected by the climate crisis. During COP21, Global Witness mentioned IPs in roughly half of their tweets, including "Stop the murder of Indigenous peoples Guardians of our forests #PaddleToParis #COP21."

[12] https://www.globalwitness.org/en/about-us/ (Accessed: June 12, 2022).

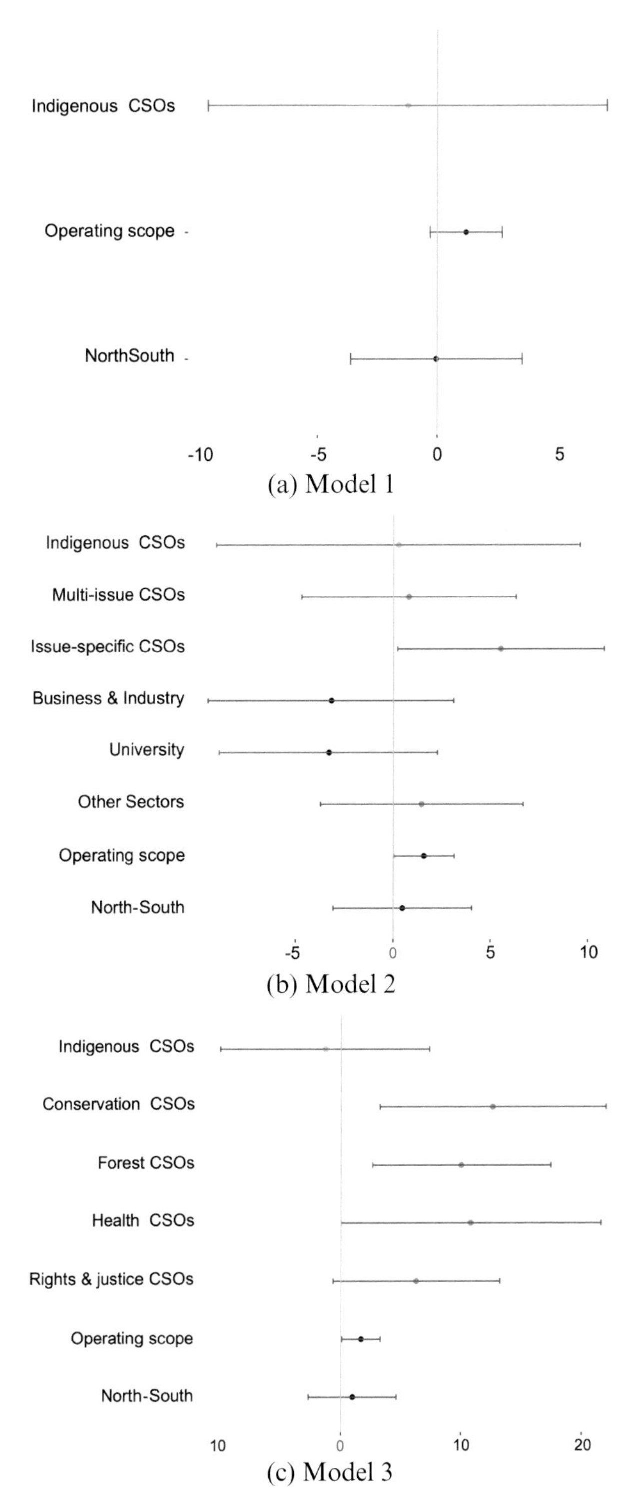

Figure 6 Average marginal effect on the number of indigenous tweets

Several CSOs focusing on forest conservation established a close connection to IPs in their advocacy discourse. Among them are FERN from Belgium,[13] Forest Peoples' Programme from the UK,[14] and California and Oregon-based NGO Pacific Forest Trust.[15] The primary goals of these organizations are the protection and sustainability of forests, to benefit climate, water, wildlife, and people's well-being. At COP21, more than half of their tweets were concerned with issues related to IPs. They highlighted the vulnerability of IPs around the world, from Cameroon to Indonesia to Brazil, advocating for a rights-based forest protection plan that includes indigenous traditional knowledge. For example, FERN tweeted, "Nice video about #indigenous Ecuadorian people fighting exploitation of their forest at #COP21 #PaddletoParis" with a link to the video.

A final example is Ford Foundation, the US-based funding giant for rights and justice-related causes. Over 65% of its tweets during COP21 adopted indigenous frames. Although initially founded as a tax scheme for the Ford family, Ford Foundation shaped our understanding of human rights over many years through its funding programs (Wong, Levi, and Deutsch, 2017). Many of the Ford Foundation's tweets reflected firm support for IPs' rights in tackling climate change. For example, it tweeted, "If we protect indigenous rights, we fight #climatechange." Several of its tweets underscored the threats posed by deforestation toward IPs.

Concerns around IPs may be a relevant area for these issue-specific CSOs, but they are not central to their missions. The findings suggest that the interest of IPs is reflected in surrogate-representation, where IPs themselves may not be spearheading to fill in the niche emerging for themselves. Instead, the indigenous discourse related to climate change is co-constructed by CSOs with a variety of issue interests.

We speculate two reasons why the IPs sector is not saturated like the women's sector. The first reason is substantive. The "surrogate-representation" pattern we observed from online data could very well be a reflection of the marginalization of IPs' participation in global climate governance in general. A substantial literature suggests that IPs' participation *inside* the UNFCCC COPs has been unequal, insufficient, and tokenized, as IPs face various obstacles at the international stage (Rashidi and Lyons, 2021; Shawoo and Thornton, 2019; Zurba and Papadopoulos, 2023). Such obstacles include a lack of financial and technical resources and the capacity to navigate a Western-dominated paradigm, which can also be translated into social media platforms. In particular, the

[13] https://www.fern.org/ (Accessed: June 12, 2022).
[14] https://www.forestpeoples.org/ (Accessed: June 12, 2022).
[15] https://www.pacificforest.org/ (Accessed: June 12, 2022).

finding suggests that IPs' organizations did not use social media as an effective tool for advocacy, although we do not know whether or not it was a conscious choice.

The second reason is methodological. Since indigenous frames are identified with co-occurrences of the seed words, such as "indigenous" and "native," they may have included false positives that are not quite indigenous tweets. Even if this is the case, however, the co-occurrence of words like "rights," "forests," and "health" reveals the nature of indigenous frames with which CSOs discuss their own issues in relation to IPs.

3.3 Conclusion

Using the idea of organizational ecology, we conceptualized CSOs into three types: multi-issue generalists, and specialists who either focus on specific issues or specific people groups. They correspond to different sets of resource base, or niches, for organizational survival. While existing research has already highlighted differences between generalist and specialist CSOs (Bush and Hadden, 2019; Eilstrup-Sangiovanni, 2019; Shibaike, 2022), we have documented further variations in specialist CSOs to represent and advocate for particular people groups. While self-representation allows CSOs to have ownership of a frame corresponding to their resource base, surrogate-representation borrows credibility from others, in which the ownership of a frame is shared.

The study of CSOs has long debated whether CSO behaviors can be best explained by resource base or principled beliefs (Cooley and Ron, 2002; Keck and Sikkink, 1998). Using COP21 as an illustrative case, we have shown that CSOs' resource base motivates their advocacy tactics, but it does not explain framing choices. We argue that the gatekeeping thesis, in which specialist CSOs take advantage of public goods provided by leading generalist CSOs, overstates the importance of resources to CSOs in general, while at the same time discounting the importance of small, but specialist, CSOs. Most CSOs are resource-poor, but this seeming disadvantage pushes them to develop their own discursive space by speaking for themselves or sharing concerns with other CSOs. Especially with the social media platforms that are readily available, our study opens up new research opportunities that can significantly enrich the understanding of CSOs' influence and agency.

With these findings from quantitative analysis, we will turn to the qualitative evidence obtained from in-depth interviews with women's CSOs and IPs organizations at COP23, COP24, and COP27. In the next section, we will draw lessons from these conversations and show how these specialist CSOs construct and shape climate change frames in multilateral venues like the COPs and use different frames to establish and expand their advocacy niche.

4 The Analysis of Women and Indigenous Peoples Advocacy

The previous section has shown different patterns of representation by CSOs when they construct a gender-climate frame and an indigenous-climate frame based on Twitter data. This section seeks to dive into how the gender and IPs niches are being filled at the UNFCCC process, especially by specialist CSOs. We argue that there are two mechanisms by which different CSOs occupy an advocacy niche. Based on our observation and interviews with CSOs involved in gender and IPs advocacy, we find that women's CSOs construct a unified gender-climate framing by exercising disciplinary power. Different CSOs provide different stories of women in relation to climate change (self-representation), but they developed a division of labor rather than fragmentation. IPs have also been working tirelessly to have their voices heard, but other specialist CSOs such as forest and agricultural CSOs also develop an indigenous-climate frame (surrogate-representation). Furthermore, we show that early-comers exercised disciplinary power over late-comers in women and gender advocacy while they did not in IPs advocacy.

With the examples of the women's CSOs and IPs at the UNFCCC, we demonstrate that competition within a niche or the lack thereof determines how CSOs adopt advocacy frames. We draw evidence from semi-structured, in-depth interviews with 24 CSOs that are dedicated to the advancement of gender equality and/or IPs in relation to climate change and policymaking at three COPs. In this section, we show how gender and IPs niches are filled and shaped by respective specialist CSOs. The gender advocacy demonstrates a pattern of disciplining in producing a unified message. The IPs advocacy contains various themes, as IPs as a group embrace diversity as a core identity. We then draw on the concept of disciplining and argue that strong leadership from the Women and Gender Constituency (WGC) at the UNFCCC developed the system in which centralized advocacy efforts are balanced by inclusive and collaborative participation. On the other hand, IPs advocacy is diverse and does not have a clear and identifiable leadership. When the niche is left open, other non-IPs CSOs can join with their own narratives. We suggest two possibilities for the future of this surrogate-representation. While this could negatively impact IPs representation and participation at the UNFCCC, it may help mainstream IPs narratives beyond the UNFCCC.

4.1 Interview and Analysis Methodology

To understand how and why CSOs adopted a gender or indigenous frame for climate change, we conducted in-depth interviews during COP27. We also complemented our analysis with the interview data we previously collected from COP23 and COP24. The interviews were pre-structured to ensure that we ask key questions about advocacy strategies and their perception of climate change advocacy (see Appendix H). Our interviews protocols have been approved by the Institutional Review Boards of Purdue University, Whitworth University, and Duke Kunshan University.

To recruit our research participants, we contacted all the CSOs participating in Side Events during the three COPs related to the theme of gender or IPs. A total of 45 organizations were contacted, and a few responded affirmatively. To increase the number of participants, we went to the COP venues. A total of 14 organizations were interviewed during COP27, 3 organizations at COP23 and 4 organizations at COP24. As we conducted interviews within the UNFCCC event areas (Blue Zones), our sample does not include CSOs that primarily engage with "outsider" strategies, such as protests outside the venue.[16] While we acknowledge the influence of "outsiders" in climate advocacy (de Moor, 2022; Fox and Brown, 1998; Hadden, 2015), our focus remains on CSOs at the UNFCCC, which continues to be a focal point for mobilization even for "outsiders" (De Moor, 2018).

We transcribed the interviews and read them manually to identify common themes. We coded and analyzed the interviews based on a grounded theory. A classical inductive methodology originally developed by Glaser and Strauss (1967), a grounded theory is a strategy for theory development without prior assumptions. We adopted an open, interactive, two-step coding process (Charmaz et al., 2012). First, we developed initial codes that described and dissected the interview data. We then constructed and assigned short, descriptive labels to texts. Finally, we compared the codes and data and used the most frequent and significant initial codes to sort, synthesize, and conceptualize all interviews. Throughout this process, we developed concepts, categories, and themes.

The grounded theory fits well for in-depth qualitative interviews, as it fosters a fluid and emergent process in data analysis. As we seek to understand CSOs' framing and advocacy strategies for climate change, it keeps us close to the empirical data, compels us to examine assumptions, and enables a deep exploration of the interviewees' experiences and beliefs. This approach prevents us from trying to interpret the data in the way that supports our theory. Instead, it helps us examine whether the data supports our theory. We repeated the same analytical process to code interviews with women's CSOs and IPs organizations. We compared the themes and concepts between the two groups of CSOs and drew conclusions on how each develops a voice in the landscape of climate governance and advocacy.

[16] COP27 was held in Egypt, in which no outside protest was allowed.

4.2 Filling the Niche: Self-Representation or Surrogate-Representation

Organizational ecology posits that competition within a population is a function of density (Carroll, 1985; Hannan and Freeman, 1989; Morin, 2020). Researchers argue that increasing competition encourages organizations to specialize in narrow niches (Aldrich and Pfeffer, 1976; Carroll, 1985; Eilstrup-Sangiovanni, 2019). Specialization can lead to the division of labor, as organizations find their unique strengths in the areas that do not overlap with one another (Eilstrup-Sangiovanni, 2019; Shibaike, 2023). Specialist organizations can also find a new niche by adopting an appropriate frame and advocacy strategies. Here, we empirically examine the processes by which CSOs fill niches at the UNFCCC.

In particular, we compare gender advocacy and IPs advocacy at the COPs and demonstrate that the two niches exhibit different dynamics through which CSOs occupy the space. Women's CSOs generate various narratives that connect gender with climate change, which demonstrates a division of labor among themselves. Women's CSOs complement each other and together forge a strong case of why gender equality matters in climate governance. By contrast, IPs' advocacy mainly comes from two camps, IPs themselves and other specialist CSOs whose expertise is in issues like conservation and health. The goal of this research is not to make a moral judgment in terms of whether such specialist CSOs joining the IPs niche are helping or harming IPs through their own expertise and programs. We are also cautious not to give a homogeneous voice to the very different indigenous peoples, nor do we want to ignore their contributions to the environment and climate. Rather, our goal is to demonstrate the diversity of narratives around IPs in climate governance. On one hand, IPs are actively lodging their claims for sovereignty, true representation, and meaningful participation at the COPs. On the other hand, specialist CSOs from other sectors, who seek indigenous participation in their programs, construct a different narrative that centers on their specialty.

4.2.1 Gender Framing: Issue Linkages and a Unified Voice

We observe a high level of specialization in women and gender advocacy at the COPs, as different CSOs seek to occupy different areas of expertise through their own messages around gender equality and climate change. Women's CSOs at the UNFCCC share a strong need to raise awareness of the connection between gender and climate change. Despite the successful mainstreaming of gender languages into the Paris Agreement and the Gender Action Plan of 2017, interviewees noticed that most people coming to COPs still did not recognize the connection between gender and the impact of climate change. One interviewee said, "You go into the climate movement and say that you're involved in this process as a feminist, often it's the first time people ever hear that gender equality and climate have something to do with each other" (Interview 1701). An interviewee from Plan International shared a similar experience when she talked to people about how child marriage had become more common during post-disasters and rising menaces of food shortage and natural resources scarcity: "They were shocked; it's evident that this is not in anyone's head here" (Interview 1801).

Women's CSOs also recognize that many social, political, and economic factors can further exacerbate gender inequality. An interviewee of the Asian-Pacific Resource and Research Center for Women (ARROW) pointed out, "The interlinkage between climate change and women's rights is just not direct. There are many other contributing factors involved, such as poverty" (Interview 1802). Another interviewee from GenderCC expressed her concern in explaining a feminist approach to understanding climate impacts (Interview 1702). According to her, ignorance stems from the deeply rooted social institutions and norms based on masculinity. The current understanding of climate change as a collective action problem and the neoliberal economic context in which climate debate takes place need to be challenged and unsettled. She said:

> There is a structural level of gender roles. It means that in our societies, the male is considered the norm. Female is just a deviation from the norm. [...] We believe that a major issue is the care economy as opposed to the so-called productive economy, and all the institutions and power behind it.

The shared sense of marginalization of gender discourse in climate governance has led to the proliferation of different frames for gender-climate issue linkage. Almost every CSO we interviewed was very intentional about explaining why a gender-sensitive perspective matters in combating climate change. As specialist CSOs, their specialty and expertise in particular areas gave them the legitimacy to fully explain the causal linkage between the impact of climate change and gender inequality. For example, one interviewee explained how water and food shortage disproportionately impacted women from resource-dependent communities (Interview 1803). As essential resources become scarce in nature, women as resource collectors are more likely to be exposed to the impact and the violence associated with the shortage. The interviewee from ARROW expanded on how negative climate impacts such as natural disasters can exacerbate women's and girls' sexual and reproductive health (Interview 1802). Another interviewee coupled the issue of urban planning and gender equality (Interview 1702). She gave an example of urban design and development to illustrate how deeply entrenched the masculine approach is in everyday life and how gender inequality and fossil fuel dependence share the same fundamental cause:

> A working man goes around in a car. He lives at home in a nice remote area. Then he goes to work, and he comes back. But what about the caregivers like women? They're so remote from the services to all these kinds of things. And this creates a lot of emissions.

Many others draw on their expertise to articulate a gender-climate story with different thematic areas, including poverty, education, investment, trade, transportation, human rights, and loss and damage. These issue areas allow specialist CSOs to garner support from their audiences, which in turn expands social resources available for women and gender advocacy. Women's CSOs are also good at identifying a new linkage and incorporate it into the broader gender advocacy. For example, an interviewee from Plan International said:

> I really do see a missing link with girls here. If they are going to be future women, and the future negotiators in these types of events, the future leaders in their communities, then unless we include them, and give them the same access to their rights as boys have, we're never going to be able to get the equality we're trying to get through these women's groups. (Interview 1801)

The variety of narratives surrounding gender-climate advocacy not only strengthens the advocacy potential but also elevates gender as an overarching theme that can thread through many problems.

The diversity of narratives is a result of increasing density in the gender advocacy niche. Increasing density gave prominence to gender advocacy in the UNFCCC, but it also compelled CSOs to demand further action for gender equality so that they can maintain the ownership of gender advocacy. Many mentioned that gender should be considered in the entire problem-solving process and in every stage of decision-making and project planning (Interview 1801; Interview 2212). An interviewee from the Centre for Feminist Foreign Policy in Berlin said, "I think if we cut back on that [demand for feminist policy], a lot of people will only meet us halfway anyway. So I think we have to strive for utopia" (Interview 2204). Similarly, a program director from Care International explained,

> We feel that not all organizations who work on gender work to the same degree that we do. They, a lot of organizations, are focused very much on the first layer of gender equality, which is about participation and inclusion, which some feminist scholars in Canada actually have referred to as

> the model of 'add women and stir.' And we don't feel that's enough. So our niche [...] is really digging into the participation side of things.

The continuous push for gender equality led to the popularity of gender frames featuring various aspects, such as poverty and education, of injustice that women face as a result of climate change. At the same time, the diversity of narratives all pointed to the same conclusion: climate action must address gender inequality.

4.2.2 IPs Framing: Increasing Density to Diverse Voices

The CSOs we interviewed represented three types of organizations that work on IPs: the first is composed of and managed by IPs themselves. For example, the Inuit Circumpolar Council (ICC) consists of Inuit from Canada, the United States, Russia, and Greenland (Denmark); the Nepal Federation of Indigenous Nations (NEFIN) represents IPs who reside in Nepal. We consider them specialist CSOs, rather than general governing bodies, as they solely focus on representing the interest of specific people groups.

The second kind is "broker" organizations that help connect IPs with other organizations, donors, and corporations. They also bring IPs to COPs and help connect IPs among themselves. Examples include Cultural Survival, a Boston-based CSO that builds capacity for IPs; Conservation International, which helps to allocate funding from the World Bank to IPs; and WWF, which partners with local communities and bring them to COPs under WWF's delegation. This second kind of CSOs is a mixture of both generalists (Conservation International, WWF) and specialists (Cultural Survival).

The final type of organizations work on an issue related to the IPs. They are often specialists who are experts on issues such as forest, conservation, or agriculture, but they seek to integrate IPs into their programs. Among our interviewees, Forest Trend and World Agroforestry are examples of this type of specialist CSOs. Based in DC, Forest Trend focuses on including indigenous communities in corporate-based carbon offset programs at COP27. World Agroforestry advocates for the inclusion of IPs in fertilizer programs. While all these CSOs draw on the issues facing IPs in the context of climate change, their narratives vary significantly more than what we observed in gender advocacy.

4.2.3 Self-Representation by IPs

The narratives produced by IPs groups are often based on their own understanding of their identity, culture, and history, instead of specific climate policy or solutions. A common theme that many IPs offered during the interviews was a critique and challenge of the Westphalian paradigm embedded in global climate governance. IPs have expressed marginalization in two ways. First, IPs do not have equal representation as the Parties, which are exclusive to nation-states in the UNFCCC process. Second, their knowledge and epistemology about nature and the environment are often sidelined or tokenized in the UNFCCC, which has been dominated by the idea that Western science is the best and only way to understand climate change. IPs thus demand the recognition of their unique status that allows them to participate in decision-making processes.

In particular, several IPs organizations commented that the category of "observer" or "non-governmental organization" is not an appropriate description about their organizations. The concept of "nongovernmental" is based on the Western, post-colonial definition of countries and national governments. To IPs, it clearly ignores the fact that IPs existed long before the establishment of Westphalian nation-state system. An interviewee from ICC Canada, representing Inuit, said:

> We do not consider ourselves a non-governmental organization. [...] Because we see indigenous people as having unique, special rights due to their relationship with colonialism. [...] We have a right to self-determination. (Interview 2209)

A representative from the Nepal Federation of Indigenous Nationalities (NEFIN) shared a similar thought, "Our organization is not an NGO. It is more than an NGO. It's actually a movement. It's a movement for equality (with other governments)" (Interview 2211). He described their status as "a sovereign nation. Sovereign in the sense that we are entitled to continue our distinct lifeways, distinct worldviews, distinct identity, culture, language, all of these." He pointed out further that the fact that they are made an observer is ironic:

> Indigenous peoples' governing systems, knowledge systems exist since way before the UN started. But what UN thinks is UN is the supreme body. But the indigenous people have those practices way before the UN was organized. So the irony is that the UN includes indigenous peoples only as an observer. Not any decision-maker or contributor. That means they are ignoring very old science.

IPs are diverse and heterogeneous groups themselves, and so their traditional approaches to nature, including concrete steps toward climate actions, are also diverse. However, they are unified in demanding equal participation, that is, participation substantively equal to nation-states in climate governance. IPs do not simply seek a legal status as Parties to the UNFCCC, but their ultimate goal is to gain recognition of indigenous rights among national governments. The representative of NEFIN pointed out that the lack of awareness persisted because there was not enough conversations or messages around IPs. IPs are acutely aware of the lack of space in international institutions, where their interest could be represented (Belfer et al., 2019). The demand for equal participation was more than just an awareness-raising campaign. The ICC actually took the initiative and drafted the "Circumpolar Inuit protocols for ethical and equitable engagement" for decision-makers as a guideline to work with Inuit (Interview 2209).

The political marginalization of IPs also meant that indigenous worldviews regarding natural resources and the environment were sidelined from the UNFCCC. The representative of NEFIN said,

> In our perspective, in our worldviews, you are an intricate part of nature. So, nature is your family. It's not a resource. It's part of your own being. You should not be overexploiting nature. You should not be exploiting your siblings, or your relatives. You should not be doing harm to your relatives. They are not resources. They cannot have monetary values. (Interview 2211)

He argues that the change in our ideas can ultimately save the planet from climate change because attitude change at such a fundamental level would make conservation initiatives and investments obsolete. From this vantage point, IPs expressed their frustration that Western science is regarded as the only reality and evidence that is applied to climate change policymaking.

A Western-centric mindset is not only found among national governments or decision-makers. Several interviewees from the IPs community expressed opposition toward market solutions to climate change such as forest carbon offsetting programs. For example, a representative from IBON was highly skeptical of the corporate interests embedded in these programs and the international institutions that manage the funding. Using an example of the REDD+ program in Kenya, he said,

> The Ogiek Indigenous peoples in Kenya are at the front lines of fighting against displacement caused by reforestation projects. [...] It's like a new form of colonizing Africa actually. Developing country resources are being used to offset the GHG emissions of developed countries. Because they are refusing to do the necessary mitigation. Without mitigation, they would rather continue business as usual and offset their pollution using resources using lands of indigenous peoples in Africa. (Interview 2208)

A representative from Cultural Survival echoed his observation:

> The corporations, as well as the governments, are undertaking projects related to the forest. The indigenous peoples have long been protecting the forest, long been used and they have guarded and protected the forest.

> The government, without giving any information, converted those forests protected by indigenous peoples into national national park conservation areas. So there has been a massive enforced displacement. (Interview 2206)

4.2.4 Surrogate-Representation from Non-indigenous CSOs

Specialist CSOs from other sectors such as forest and agriculture have also been actively engaged with IPs. These specialist CSOs emphasize the role of IPs and their contributions in tackling climate change, while acknowledging the difficulties of IPs participation in climate governance. For example, a representative from Forest Trend said,

> I think we need to be calling the indigenous people the main actors of this because they have demonstrated that they know how to conserve, they know how to stop deforestation, and they have been taking care of the ecosystem for decades. But sometimes they don't have a place to be here. Governments don't have a dialogue with indigenous peoples. And companies are also the same. [...] Indigenous peoples need to be in the center because they are protecting the ecosystem, and they have the knowledge to continue doing that. (Interview 2207)

However, being experts of specific issues also means that their narratives are centered around the programs that they promote. Their relationship with the IPs community can be intricate, as it could be seen as a form of cooptation. Specialist CSOs carefully emphasize that their work is capacity-building for IPs and that they are not taking agency away from IPs.

Specialist CSOs also recognize the knowledge of IPs concerning nature conservation and sustainable agriculture. Another example comes from World Agroforestry, a specialist CSO working to ensure food security and environmental sustainability.[17] The organization's representative discussed its fertilizer program and his plan to work with IPs in Africa. The organization is aware of the IPs' various approaches to agriculture. The interviewee said, "The knowledge that indigenous communities have is also important. As you can imagine, farmers' knowledge comes from many years of observation and it has been transmitted through many generations. So they have been running the longest experiment than anybody can imagine." However, rather than following the IPs' approaches to agriculture, he argues, "we need to understand how [indigenous] knowledge supplements and compliments the knowledge that is technical" (Interview 2213).

Well aware of the pushback on using fertilizers in agriculture, he argued that access to fertilizers was an equity issue. He mentioned that the organization's

[17] World Agroforestry. www.worldagroforestry.org (Accessed: June 14, 2023).

program would achieve both improved agricultural productivity for the local community and increased carbon sink for the world. To further support the argument, he connected the program with a rights-based approach to IPs:

> Communities have rights. This could be rights to their land. This could be rights to the indigenous local knowledge. This could be fundamental rights to freedoms, to express their cultures and all of that, but also just the right to live and the right to food and shelter and respect and all of that. And I think every effort that supports climate mitigation, climate adaptation should be sensitive to those rights.

Certainly, the organization uses its expertise and the scientific languages, such as "nitrogen fixation," "mechanisms of runoff," and "leaching," which we rarely hear from IPs in their narrative. These CSOs have a strong desire to work with local indigenous communities, but whether IPs will welcome their initiatives is an open question.

4.3 Niche-Filling and the Disciplinary Process

In Section 2, we provided a theoretical argument about disciplinary power in the process of niche-filling. The characteristics of early-comers that occupy a given niche shape the outlook of opportunities for other organizations (Lake, 2021; Morin, 2020). We find strong evidence of early-comer advantage in the women and gender niche at the COPs, in which those who engaged with the frame earlier on exercised the power to shape appropriate forms of advocacy for gender-climate issues. By contrast, the IP community is still at the early stage of niche-filling, and IP organizations have not policed the narratives about themselves. As different types of specialist CSOs are scrambling to frame IPs differently, it is important to examine who might dominate the niche, which can determine the long-term prospect of how IPs will be perceived at the UNFCCC process.

4.3.1 Women and Gender: Leadership and Discipline

The emergence of gender frames at the UNFCCC was credited to a core group of specialist CSOs with a focus on women's rights and gender equality. CSOs like WEDO, GenderCC, WCEF, and APWLD have taken leadership during the early days when a gender perspective first started to surface at the UNFCCC (Interview 1702). These CSOs were also mostly from the global North and thus able to bring funding with them to the UNFCCC process. They became magnets that drew more CSOs into the space (Interview 1702). From the organizational ecology perspective, these early-comer CSOs not only developed a new niche for other CSOs with similar issue foci but also expanded the niche by providing crucial material and social resources.

What they also brought to the UNFCCC was their previous advocacy experience in other multilateral institutions. Talking about the formation of WGC, an interviewee commented that there was a "big step forward when a number of women's groups who had before attended the General Sustainability process in the Major Groups process joined the UNFCCC process" (Interview 1702). They built upon existing collaborative relations and carried over the knowledge of how to engage in the multilateral process. Another interviewee also mentioned a similar process of knowledge accumulation outside of the UNFCCC, mostly through the Global Gender and Climate Alliance and Women's Major Groups that engaged in the Sustainable Development Goals process, public health, and biodiversity. Their previous experience in multilateral processes helped them to navigate a new space at the UNFCCC (Interview 1703).

Over the years, the early-comer CSOs accumulated political knowledge of how to navigate the UNFCCC process. When talking about the institutional opportunities provided by the UNFCCC, one interviewee said:

> I think when it comes to civil society being able to effectively engage in a process, one of the key things is if there is an institutionalized framework for that, the fact that the UNFCCC has constituencies, it means that there is a need to continually be engaged. We [referring to the WGC] have morning meetings. If everything else stopped, the morning meetings is what we would always have. (Interview 1703)

Their political experience enabled them to produce reports and documents that would engage with policymakers. A representative from Plan International Canada said,

> We are able to not only highlight the challenges but also the solutions with clear, actionable insights and policy recommendations, which does make our advocacy more effective. And we were able to use that as a launching pad for conversations for parliamentarians and the minister of environment and climate change (of Canada). (Interview 2212)

The early-comer CSOs also see the need to *discipline* the women and gender advocacy community. By disciplining, we mean the efforts of early-comer CSOs to shape and unify gender-climate narratives and help other CSOs to learn the norms and rules of engagement in the UNFCCC process. The WGC has always been open for newcomers and collaborations, as an interviewee said, "We try to provide the space and environment capacity and resources to have an open caucus. We try to host a space where everyone can come and share, that is really a space for co-learning and for movement building" (Interview 1703).

At the same time, early-comer CSOs continue to lead the process of shaping the key demands from the WGC. In short, any CSOs are welcomed to join the gender-climate advocacy community, but they need to follow the footsteps of the early-comer CSOs.

The early-comer CSOs do recognize the need to have a centralized voice for the entire WGC. One interviewee said that "It's impossible to work as one organization to follow all aspects of the process," and "collective advocacy is necessary to be able to have an influence" (Interview 1703). Another also commented, "For us to be truly effective, we need multiple organizations speaking from the same songbook and pushing for the same agenda" (Interview 2212). While an increasing number of women's CSOs have joined the WGC over the years, only a few are actively involved in strategizing for the entire WGC: "Within the constituency, what makes it effective for advocacy, you have to have a small, dedicated team of individuals who are following the process, who are building relationships with governments" (Interview 1703). The centralized approach is reflected in the list of "key demands" that the WGC put forward at every COP. In an increasingly populated space, this is made possible only when there is a small number of CSOs intentionally shaping the list. They do not necessarily "gatekeep" what is included and what is not, but by providing guidelines and perimeters, they ensure that their collective advocacy would stay focused and consistent throughout the UNFCCC process.

At the same time, there are inevitable drawbacks when a few CSOs try to discipline the entire gender-climate advocacy community. In particular, Southern CSOs often feel pressured into the practices of Northern CSOs and their values. While Northern CSOs were the first ones that started to establish women's advocacy space at the UNFCCC, Southern CSOs felt excluded during the early years. Between 2007 and 2008, when the women's CSOs gathered together at the UNFCCC to plan for institutionalized participation and bolster their political influence, some of those Southern CSOs were excluded from the initial caucus where CSOs deliberated participation strategies (Nagel, 2015). Southern CSOs still gathered with their Northern peers at the COPs, but the exclusion from major decision-makings was felt keenly.

The North-South divide is further complicated when the Northern CSOs are usually the donors or help connect their Southern partners with donors. Northern CSOs often explicitly demand certain characteristics for local partner organizations. ADRA (Adventist Development and Relief Agency), an international humanitarian organization, has a large number of local partners worldwide to implement gender-related programs such as education for women. This screening process is a top-down approach, where it is ADRA that chooses its partners rather than local partners coming to work with ADRA.

Program directors from ADRA Canada explained how those partners are selected:

> What happens is that when we go to a country, we always go with an ADRA office that's registered in that particular country. They are the ones who serve as the focal persons that connect with the local experts. [...] They go through some kind of criteria of looking at the history of this NGO (Local women's NGO) as well. What's their reputation, what kind of funding are they able to account for. (Interview 2202)

Similarly, a program director from Care International explained their partnership with local organizations. He emphasizes that Care International is moving toward decolonization and genuine partnership with local organizations. At the same time, he admitted, "We used to take donor money, and we would subcontract with the organizations to implement projects that, fundamentally were designed either by the donor or us" (Interview 2203).

At times, CSOs find themselves in a difficult situation where they have a mandate from donors to ensure certain gender quotas for local projects but local communities are not ready for women's leadership. A representative from Conservation International mentioned:

> Sometimes the [World] Bank and the big donors will say, you need to put 50% women and 50% men. But then, it's challenging because the Bank comes with their rules, and their rules are pretty much centered in West. In the indigenous peoples' caucus, they have been really fighting for with the big donor. Like, you're putting all of us in one bag, and then you're telling us that we have to do things certain way. (Interview 2205)

Finally, the disciplinary process is also reflected in how experienced CSOs from the global North helped newcomers to socialize into the UNFCCC process. Several interviewees mentioned a major challenge, that is to navigate the multilateral space, including the knowledge of the process and the ability to speak the "acceptable language" at the COP. One interviewee commented:

> There are some very, very formal ways of engaging [at the UNFCCC] as an observer and it can be incredibly intimidating. It is very exclusive. I think it really excludes young people and activists from the global South and indigenous people that just simply don't speak that diplomatic international negotiation language. [...] And I think if you are not able to bring your arguments in that kind of language, they will not be heard. Your opinion will be marginalized because you don't fit the mainstream experience, the mainstream way of talking about issues. (Interview 1701)

Another interviewee also made the observation that, "at that time [when the WGC was established], there were a lot of training, which we all attended, and which gave us a whole lot of knowledge about how to address gender and climate change and the kind of activism we can start doing from our own" (Interview 1803). Therefore, the early-comer CSOs of the WGC not only bring Southern women to the UNFCCC but help them translate their stories into the language of the multilateral processes and give them skills and tactics to become effective advocates of gender equality and women's rights (Interview 1701; Interview 1702; Interview 1805).

Tracing the establishment and shaping of the gender-climate niche at the UNFCCC, we observe that the niche has developed a pattern through an experienced leadership and their disciplinary efforts. The group of CSOs that had carved out the niche brought strong leadership and provided knowledge and experience to navigate the space for those who followed. Those CSOs were able to set the tone of collaboration, professional engagement, and continuity over the years of UNFCCC process.

4.3.2 Indigenous Peoples: Diverse Voices

IPs have been present in the UNFCCC process for the past twenty-five years. Despite establishing platforms for caucusing and sharing information, IPs advocacy at the UNFCCC still experiences numerous barriers to participation. Formal acknowledgment does not automatically translate into meaningful participation and policy outcomes for IPs. In our research, we move beyond the common critique that IPs are "marginalized" in the UNFCCC process and feature the agency of IPs. Based on the strategies and core beliefs of IPs organizations, we find that the indigenous-climate niche is open to other CSOs who can shape and produce advocacy discourse on IPs. If IPs do not (or cannot) discipline the narratives about themselves, who is filling the space? While we are unable to project with certainty what might happen in the future of IPs advocacy at the UNFCCC, we highlight some of the implications around the current state of IPs advocacy and what it could mean to IPs as well as other CSOs.

Multiple interviewees have mentioned the barriers for IPs to meaningfully engage at the UNFCCC space (Interview 2206; Interview 2208; Interview 2214). The literature on IPs and climate governance has also pointed out the issue of marginalization (Comberti et al., 2019). Politically, IPs are recognized as non-Party *observers* without decision-making power in the negotiation processes (Belfer et al., 2019; Comberti et al., 2019). Lumping them into one big constituency of IPs also erases the diversity and heterogeneity of IPs themselves (Belfer et al., 2019). Different IPs organizations have different priorities and capacities, which could sometimes lead to disagreements and conflicts among the IPs community (Zurba and Papadopoulos, 2023).

Economically, IPs generally suffer from the lack of resources to attend COPs. This has been a challenge mentioned by our interviewees as well. In the broader context, IPs worldwide have experienced colonial subjugation and destruction of their culture, peoples, and lands. They were forced out of their lands and often resided in remote places with adverse climate conditions. Comberti et al. (2019) use the example of the Navajo Nation in the United States, where their reserve is in the areas that recorded the highest temperature increase in the country.

The lack of political and economic capacities among IPs' organizations leaves the space for surrogate-representation. CSOs have mentioned the lack of capacities in the context of IPs frequently during our interviews. For example, an interviewee from Conservation International mentioned, "What we found out was that there were not too many indigenous organizations had the capacity to manage that amount of funds" (Interview 2205). An interviewee from Forest Trend also acknowledged the need for capacity-building:

> [IPs] need support, capacity building. They need to work always in an alliance in order to be ready to work with markets, and would be ready to work with large scale foundations or international cooperation. And the role of Forest Trend is to provide this support to indigenous peoples. (Interview 2207)

Epistemologically, IPs often hold a different set of worldviews about nature that are not commensurate with Western science (Interview 2211). While there is a great diversity of views among IPs too, they often give nature significant spiritual and relational meaning instead of treating them as resources for use. IPs manage the natural environment according to their traditional beliefs, which could be in a fundamental disagreement with Western science and knowledge. When the interpretation of reality has been sidelined for a long time, it is difficult for IPs to have a meaningful dialogue in the venue based on the diplomatic norms and understandings that had developed with a history of colonialism.

Fundamentally different views about nature also prevents IPs from disciplining advocacy discourse among non-IPs CSOs; instead, conflict and disagreement may emerge. For example, Inuit hunt wildlife for subsistence. To Inuit, the hunting of wildlife, such as seals and caribou, is an integral part of their culture that has continued from the origin of their history. An interviewee from the ICC Canada mentioned that some environmental CSOs were "too radical" and that animal rights movement has led to the trade ban on seal skin products in Europe and the United States (Interview 2209). This has had a direct damage to the livelihood of Inuit and their relationship with wildlife.

The interviewee also commented on the double standard of these environmental movements. As Inuit butcher animals on the ice, their photos can easily be repulsive to people who live in urban areas. But "the Western world is depending on butchers as well to eat cows, pigs, ducks, fish, all kinds of animals. But the Western world doesn't see the butchering happening and what it looks like and all the blood that comes out of it" (Interview 2209). Although both IPs and other environmental CSOs are considered pro-environment on the surface, the deeper philosophical basis differs significantly between IPs and the environmental movements developed in the West. While Western scientific knowledge and principles about environmental protection are prioritized at the UNFCCC, the voices of IPs continue to be sidelined as anecdotal (Comberti et al., 2019).

IPs generally do not, or cannot, discipline the narratives against traditional practices. Instead, it is a time-consuming process of demanding their own rights. The interviewee acknowledged that there had been some progress in terms of how non-IPs CSOs understand IPs' traditional practices:

> If their philosophies conflict with ours or contradict ours, we cannot work with non-governmental organizations (NGOs). Certain NGOs we do, and NGOs are starting to gain a much better understanding because we had to speak out and advocate for our rights. (Interview 2209)

She then noted that the major CSO that contributed to the ban on seal trade in Europe and the United States made an official apology in 2014 for their misguided campaigns in the 1970s–80s.

4.3.3 The Leadership Void and Possible Consequences

Unlike the WGC, where there is a clear leadership structure, the IPs' caucus at the UNFCCC is more loosely structured. No group of CSOs discipline advocacy discourse or strategies for IPs. Leadership may not be a necessary condition for successful advocacy, but from the perspective of organizational ecology, early-comers create a frame that carves out their niche, and latecomers will emulate or advance the framing effort to populate it. By generating a unified voice, CSOs expand mutually beneficial collaborative opportunities (Allan and Hadden, 2017).

We find that the absence of a disciplinary process in IPs advocacy has to do with the broader structural conditions in which IPs are embedded. Unlike women and gender advocacy, the multilateral institutions that recognize IPs organizations are still limited. For example, the ICC became the first IPs organization to receive a *provisional* consultative member of the International

Maritime Organization in 2021. However, the representative of the ICC noted its limitation:

> That's not recognized as an indigenous people's organization because they consultative status members of the IMO. And I've said, it should be recognized as an indigenous people's organization. (Interview 2209)

While the UN Permanent Forum on Indigenous Issues established in 2000 provides a useful model for IPs participation in global governance, it has not spread broadly across multilateral institutions. The lack of participatory arrangements in global governance prevents IPs from establishing a standard of appropriate languages and behaviors for other CSOs when they engage in IPs advocacy.

The diversity of IPs worldviews has also contributed to the diversity of narratives around IPs. To be sure, the diversity of IPs still lacks wide recognition at the UNFCCC. One of the obstacles that IPs face at the UNFCCC that they are often treated as a homogeneous group of *people* rather than diverse *peoples*. However, because diversity is a core belief shared among IPs (as evident in their approaches to nature conservation from the Amazon to the Antarctic), their climate frames also vary widely. This in turn makes it difficult to create a unified frame for the IP community as a whole beyond their demand for participation in the UNFCCC process.

Another factor is that IPs have historically faced exploitation by market-oriented actors. Initiatives like REDD+ are still contested among IPs, and there is no consensus as to whether they will be beneficial to IPs' right to self-determination, one of the few claims that have a consensus among IPs and their allies. IPs are also very careful in choosing CSO collaborators due to the epistemological differences discussed previously. Although there are institutional mechanisms to utilize the lands of IPs for the purpose of climate change mitigation, the history of exploitation by market actors prevents IPs to unanimously support such initiatives or CSOs, and in reality, nobody knows whether such market actors can sufficiently reflect the interests of IPs.

We highlight what diverse voices for IPs could mean to the future of IPs advocacy. First, it can be that the niche for IPs advocacy will be left open for non-IPs CSOs to fill in and possibly dominate. As discussed earlier, the UNFCCC is an intergovernmental venue where nation-states are the primary participants with ultimate decision-making power. This is one issue that IPs and their allies have a consensus: IPs should be allowed to participate in decision-making processes (Interview 2209; Interview 2211). Because the issue of participation is closely tied to the right to self-determination, specific issues concerning the use and protection of IPs' lands, for example, may remain as a secondary issue. As such, unless IPs are given the status similar to Parties to the UNFCCC, it may be difficult to find a unified voice. Instead, CSOs may borrow, or even appropriate, IPs advocacy to advance their positions at the UNFCCC.

Second, the prioritization of Western scientific knowledge at the UNFCCC would disproportionately favor those who can speak the same language – not just English but also technical terms for science and diplomacy. Even though there has been a movement toward acknowledging IPs' knowledge in scientific assessments and, more specifically, within the UNFCCC Subsidiary Body for Scientific and Technological Advice (Ford et al., 2016; Maldonado et al., 2016; Tormos-Aponte, 2021), historical discrimination and the lasting impact of colonialism may make it difficult to mainstream IPs' approaches to nature and climate change. By contrast, CSOs with scientific approaches may be able to shape IPs advocacy, as they can use the language that can easily appeal to Western audiences.

Finally, who exercise disciplinary power can shape the future of IPs advocacy at the UNFCCC. If, for example, conservation CSOs can dominate IPs advocacy by highlighting the aspects of carbon emissions reduction through IPs' forests, other concerns facing IPs could be sidelined. To be sure, no CSOs at the UNFCCC would argue that other concerns, such as non-economic loss and damage to IPs, are of little importance, but given the scarcity of resources for the IPs community, some issues inevitably gain more attention and support than other issues. The current state of IPs advocacy is inclusive of various policy positions, but if it changes in the future, it could potentially hinder solidarity among IPs.

4.4 Conclusion

Based on the theory of organizational ecology, this section has illustrated different patterns by which CSOs fill in their advocacy niches. While gender advocacy is populated by women's CSO representing their own interests, IPs advocacy is spearheaded by IPs organizations, IPs advocacy CSOs, and non-IPs CSOs specializing in environmental issues, such as forest conservation and sustainable agriculture. We have argued that the main difference between these niche-filling mechanisms hinges on whether early-comer CSOs in the advocacy niche discipline advocacy discourse and tactics for late-comer CSOs. The early-comer CSOs among the WGC were not as dismissive of late-comer CSOs as the word "gatekeepers" suggests, but they did ensure that members have the capacity to follow the UNFCCC process and agreements on key issues at each COP. In IPs advocacy, we did not observe a strong disciplinary process. IPs themselves are diverse and embrace various approaches to climate mitigation and nature conservation at a deep philosophical level, which in turn created opportunities for CSOs to incorporate IPs in their advocacy strategies.

We are not certain about the consequences of these different niche-filling mechanisms for the future of climate advocacy, but we highlight a few important cautionary factors. One is that the absence of the disciplinary process can open up for opportunistic behaviors and tokenism, which weakens the solidarity of an advocacy community as a whole. In the case of IPs advocacy, there has already been a divisive issue like REDD+, and the lack of CSOs that can exercise disciplinary power could have a long-term negative impact. Another possibility is the opposite effect, that is, exercising disciplinary power may have a negative impact on the advocacy community. There is an increasing sense of disappointment at the UNFCCC process, and CSOs like Fridays for Future are taking alternative actions outside traditional venues (de Moor, 2022). A stronger disciplinary process may force more CSOs to move away from the UNFCCC and seek out alternative sites. In sum, we call for further research on the trajectories of gender and IPs advocacy at the UNFCCC.

5 Conclusion

The UNFCCC has been a central venue for climate change since its inception. A large number of CSOs, along with other non-state and state actors, have participated in the UNFCCC process to shape the understanding around climate change. As we have demonstrated in previous sections, CSOs construct narratives about climate change by using various frames. Drawing on the literature on advocacy framing and the theoretical framework of organizational ecology, we have argued that CSOs choose particular frames in order to gain social and material resources from their audiences. In particular, we have shown that specialist CSOs – groups that target a narrow set of issues or people groups – have two different ways by which they occupy their resource space (i.e. niches).

Both quantitatively and qualitatively, we have shown that women's CSOs are *self-representative* in the sense that they drive gender-climate advocacy at the UNFCCC. The core group of CSOs exercise disciplinarity for the gender advocacy community as a whole by providing expertise and teaching appropriate forms of advocacy at COPs through the WGC. By contrast, while different IPs groups are highly active at the UNFCCC venue, IPs advocacy is also characterized by *surrogate-representative*, as a diverse group of CSOs advocate for IPs in various ways. The diverse voices for IPs come from the fact that IPs organizations do not (or cannot) exercise disciplinary power over other CSOs talking about IPs in their own programs.

5.1 Theoretical Implications

Existing research has shown how CSOs and their framing efforts might impact policy processes. this study contributes to our understanding of how CSOs adopt different frames and create a common understanding what climate change is really about. The literature on framing in global environmental governance shows that fit with existing norms helped establish dominant ideas (Bernstein, 2001; Epstein, 2008). While we agree that normative backgrounds are important to how CSOs adopt frames, especially the disciplinary process of frame development, we also find that it is difficult to a priori predict which norms are most important to the success of a particular frame. Trying to avoid the risk of providing ad hoc explanations, our approach here gives us a sense of what might happen next given the current state of climate advocacy. For example, we provided two possibilities that IPs' advocacy can be appropriated by issue-specific CSOs or that the diversity of CSOs advocating for IPs can mainstream their claims. We cannot predict the outcome with certainty, but ecological dynamics among CSOs tell us how certain frames are sidelined or popularized based on the characteristics of organizational populations in a given advocacy space.

Looking at organizational characteristics, the literature on transnational advocacy has focused on the power of "gatekeepers" that allow certain ideas to spread while blocking others (Bob, 2005; Carpenter, 2007). In the context of the UNFCCC, Allan (2020) finds that well-connected NGOs use frames to broaden issue-based coalitions and raise issue salience. While we also observed that prominent CSOs like GenderCC and WEDO shaped the overall tone of gender advocacy at the UNFCCC, the process was aimed at the capacity-building of late-comers (hence, we call it disciplining instead of gatekeeping). Their role was also to consider an overarching frame for the entire community, rather than to evaluate whether or not certain groups or claims should be a part of their advocacy discourse.

With the expansion of online space, the findings suggest that a group of dedicated, small CSOs may be able to overcome gatekeepers in deciding how climate change should be framed. To be sure, this study did not directly measure the size of CSOs, so we do not know the exact scale of specialist CSOs beyond the ones that we looked closely in Section 4. However, as specialist CSOs rely on a narrow niche, they tend to be small in terms of both social and economic resources and therefore are unlikely to influence policy processes individually. This study joins the recent effort to examine populations of CSOs (as opposed to individual CSOs) and evaluate their power and strategies in global climate governance (Allan, 2020; Hadden, 2015). Our contribution is novel in the sense that, instead of networks of CSOs, how CSOs occupy their niche (i.e. how competition unfolds among them) can explain their framing choices.

This study also advances our understanding of ecological dynamics among CSOs not only at the UNFCCC but also in global governance more generally. Existing research has shown that CSOs specialize in narrow niches, such as issues and tactics, to survive as well as to expand resource availability for their organizational populations (Bush and Hadden, 2019; Eilstrup-Sangiovanni, 2019; Shibaike, 2023). We have provided theoretical and empirical accounts for the mechanisms of specialization using the UNFCCC as an illustrating example. It is true, as the literature suggests, that market competition encourages specialization, but specialization is not a uniform phenomenon across different organizational populations. Depending on the characteristics of early-comers, the evolution of a niche looks quite different from one to another, as we have shown in the cases of gender and IP advocacy niches. The findings provide us new ways to look at existing research on organizational ecology in the context of global governance, not necessarily to challenge the existing findings but to unpack what has been treated as a dichotomous phenomena between generalist and specialist organizations (Abbott, Green, and Keohane, 2016).

Finally, this study provided a new way to think about advocacy discourse at the UNFCCC. Existing research on civil society participation at the UNFCCC often focuses on the constituency membership to examine the representation of various interests in climate advocacy (Allan, 2020; Cabré, 2011). Our approach allows us to capture both organizational identities and framing choices without reducing one to the other. This is important because organizational identities correspond with framing choices in some cases (e.g. gender advocacy) but not in others (e.g. IPs advocacy). Our theoretical framework can therefore be applied to other intergovernmental venues beyond the UNFCCC. As intergovernmental venues increase their openness to CSOs and other non-state actors (Tallberg, Sommerer, and Squatrito, 2013), the complexity of participants also increases in terms of organizational identities. To assess whether civil society participation produces meaningful outcomes – be it international agreements or normative changes – we need to take both organizational characteristics and discursive strategies into account.

5.2 Future Research

The future of climate advocacy, especially for IPs, is a question to be addressed in future research. With the rise of climate justice advocacy at the UNFCCC, the salience of IPs in climate governance has increased substantially over the past decade. As we have shown in previous sections, however, the popularity of indigenous frames does not necessarily mean the advancement of IPs' status in climate governance. For example, PES programs for resource management and nature conservation are still contested among IPs with the concern that colonial actors like states and corporations will have power to control the IPs' lands (Claeys and Delgado Pugley, 2017; Smith et al., 2019). Although the diversity of voices in IPs' advocacy may help bolster the standing of IPs at the UNFCCC, future research needs to continue to pay attention to the risk of tokenism and the perils of neocolonial takeover of indigenous lands.

By contrast, we found that gender-climate advocacy was more centralized than IPs advocacy, led by a core group of CSOs at the WGC. There is a shared sense of progress among CSOs trying to do better than what states or even they themselves used to do in the past. Our argument is not that women's advocacy has been successful but that it has developed differently from IPs advocacy, as there are many structural conditions outside climate advocacy that established gender discourse (Hafner-Burton and Pollack, 2002). In fact, the advancement of gender advocacy and the requirement of women's leadership quotas can sometimes be at odds with what local communities can accept today. We suggest that this tension between international women's rights and local norms in the context of climate finance is an important area of research for both practitioners and academics.

Recently, the UNFCCC as a whole has been challenged as a legitimate venue to discuss climate goals (Zhao, 2023). As member states continue to fail to establish rigorous emission reduction measures, CSOs like Fridays for Future are moving away from the UNFCCC process and running independent campaigns globally (de Moor, 2022). For example, Greta Thunberg, the famous youth activist from Sweden, has repeatedly accused COP meetings as greenwashing opportunities for wealthy states and corporations.[18] Although the rise of justice discourse at the UNFCCC was driven by radical CSOs (Bäckstrand and Lövbrand, 2006; Hadden, 2015), these CSOs may be the first group to turn away from the UNFCCC process. In the context of this study, it means that the niche for CSOs adopting justice frames will shrink, thereby intensifying competition for social and material resources among CSOs advocating for gender and IPs alike. If this is indeed the case, future research may need to explore both inside and outside of the UNFCCC process to understand the dynamics of climate frame adoption among CSOs.

Finally, while this study has explored how CSOs develop climate justice discourse, it did not examine the relationship between climate frames and

[18] https://www.theguardian.com/environment/2022/oct/31/greta-thunberg-to-skip-greenwashing-cop27-climate-summit-in-egypt (Accessed: October, 20, 2023).

policy outcomes. Although the analysis of policy outcomes is an important area of study, formal agreements and initiatives are not the *only* outcome of climate advocacy. We argue that the intersubjective meaning of "climate change" is in and of itself an important outcome of CSO advocacy, as it generates a focal point of policy coordination and contention among various stakeholders. In fact, existing research has also suggested that framing can influence how Parties resolve differences in policy positions and how CSOs build coalitions at the UNFCCC (Allan and Hadden, 2017; Vanhala and Hestbaek, 2016). However, the extent to which CSO advocacy might influence intergovernmental processes remains an open question. Many practitioners in our interview research have also expressed concerns that the dual process at the UNFCCC between state and non-state actors may become even further apart in future COPs. In addition to the interorganizational dynamics investigated in this study, future research should examine whether and how climate discourse developed by CSOs travels to the languages of government delegates and official agreements.

Appendix

A Description of NGO Sectors

Table A1 Full list of NGO sectors

Nongovernmental Organizations (NGO) sectors	Description	Number of CSOs
Built environment	CSOs dealing with cities, urban systems and planning	17
Business/industry	Business and industry CSOs. Representation of businesses, business associations. Not included those specifically under another category, such as energy or forests	67
Climate change	CSOs whose raison d'etre is addressing multiple issues in climate change	51
Conservation	CSOs dealing with biodiversity, conservation of natural resources and area	15
Development	CSOs whose main goal is development, hunger and poverty reduction, or human development.	34
Education/capacity building	CSOs other than universities whose primary goal is education or capacity building	23
Energy	CSOs dealing with energy issues	40
Environment	CSOs dealing with multiple environmental issues, broadly defined. Not including environmental CSOs engaging in single issues such as forest, agriculture, etc.	91

Table A1 (Cont.)

Nongovernmental Organizations (NGO) sectors	Description	Number of CSOs
Finance/market mechanism	CSOs engaging in the design, management, and monitoring of carbon trade, carbon offset programs, and other market-oriented mechanism. Climate financing agencies	12
Food, soils, and agriculture	CSOs dealing with food production, agriculture and soil degradation	
Forest	CSOs dealing primarily with forests	28
Health	CSOs dedicated to health issues	11
Indigenous peoples	Organizations and groups dealing with indigenous peoples issues or whose constituency consists primarily of indigenous peoples	21
Legal practice	Lawyer, legal and law-related CSOs	6
Other/unknown	CSOs that do not at under any other category or whose mission could not be determined.	56
Pollution/waste	CSOs dealing with air pollution and solid wastes	3
Religious/spiritual	Faith-based CSOs.	25
Rights/justice	CSOs based on a rights and/or a social justice approach. CSOs with rights-category such as women or indigenous peoples not included.	34

Table A1 (Cont.)

Nongovernmental Organizations (NGO) sectors	Description	Number of CSOs
Science/engineering	Scientific and engineering CSOs, excluding universities, not primarily involved in any of the other categories.	33
Sustainable development	CSOs whose stated main goal is sustainable development or both environment/sustainability and development	19
Think tank	Think tanks and CSOs focusing on policy and/or international relations	10
Transport	Transport related CSOs	7
University	Higher education institutions.	140
Water, oceans, and fisheries	CSOs addressing oceans, fresh water and fisheries issues	14
Women	CSOs primarily concerned with women's rights and issues	13
Youth/children	CSOs addressing youth and children issues, or with youth as their primary constituency.	36

B Boxplot of Frame Adoption Rate

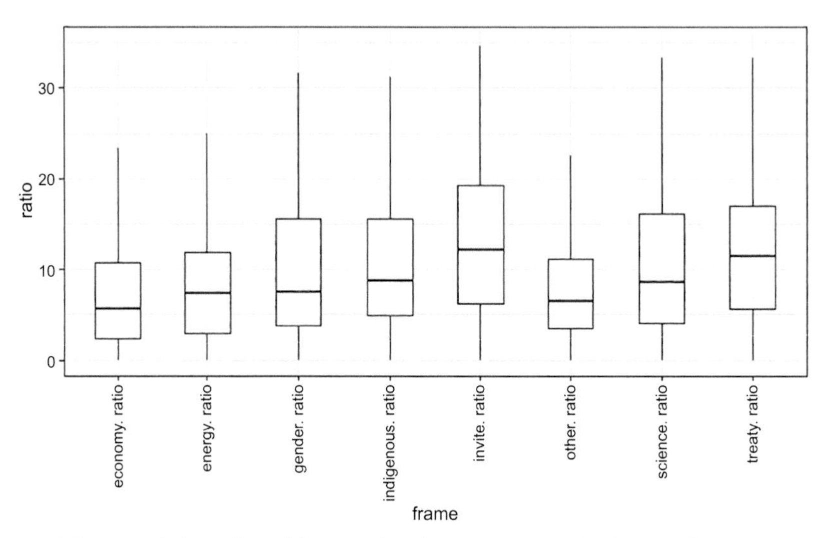

Figure B1 Boxplot of frame adoption rate across CSOs. Outliers are removed for visualization

C OLS for Frame Adoption Extremity

Table C1 Result of OLS analysis for frame choice extremity

	Dependent variable:
	Frame choice extremity
Issue-specific CSOs	7.820***
	(2.036)
People-specific CSOs	4.926
	(2.806)
Other CSOs	4.213*
	(1.934)
National	−0.827
	(1.892)
Regional	−2.480
	(2.430)
South	3.088
	(2.908)
Constant	27.797***
	(1.644)
Observations	231
R^2	0.067
Adjusted R^2	0.042
Residual Std. Error	10.908 (df = 224)
F Statistic	2.697* (df = 6; 224)

Note: *$p<0.05$; **$p<0.01$; ***$p<0.001$

D Baseline Model for All Tweets during COP21 (Multi-issue CSOs as Reference Category)

Table D1 Results of baseline model for all tweets during COP21

	Dependent variable:
	Tweets by CSO observers at COP21
Issue-specific CSOs	−0.834***
	(0.284)
People-specific CSOs	−0.828**
	(0.374)
Other CSOs	−1.705***
	(0.248)
Operating scope	0.222**
	(0.106)
Global North	1.099***
	(0.253)
Constant	3.028***
	(0.282)
Observations	786
Log likelihood	−2,476.000
Theta	0.167***
Akaile Inf. Crit.	4,964.000

Note: *p<0.05; **p<0.01; ***p<0.001

E Negative Binomial Regression Output for All Tweets during COP21

Table E1 Results of binomial regression model for all tweets during COP21

	Dependent variable:	
	Number of tweets by CSO observers at COP21	
	(1)	**(2)**
Multi-issue CSOs	2.502***	1.722***
	(0.745)	(0.244)
Issue-specific CSOs	1.694 **	
	(0.738)	
People-specific CSOs	1.668 **	
	(0.776)	
Business/Industry	1.638 **	
	(0.774)	
University	−1.152	
	(0.749)	
Others	−0.979	
	(0.735)	
Built environment		1.976***
		(0.600)
Conservation		0.516
		(0.638)
Finance/market mechanism		1.254*
		(0.741)
Food/Agriculture		0.595
		(0.620)
Forest		0.400
		(0.493)
Health		0.626
		(0.742)
Legal		0.873
		(1.091)
Pollution/waste		−1.652
		(1.444)

Table E1 (Cont.)

	Dependent variable:	
	Number of tweets by CSO observers at COP21	
	(1)	**(2)**
Rights/justice		1.008**
		(0.457)
Sustainable development		0.458
		(0.588)
Transport		0.986
		(0.924)
Indigenous peoples		−0.099
		(0.595)
Women		1.323*
		(0.744)
Youth/Children		1.062**
		(0.433)
Operation scope	0.368***	0.266**
	(0.107)	(0.106)
Global North	1.182***	1.043***
	(0.242)	(0.258)
Constant	0.260	1.293***
	(0.746)	(0.258)
Observations	786	786
Log likelihood	−2,445.651	−2,468.733
Theta	0.184***(0.011)	0.171***(0.010)
Akaile Inf. Crit.	4,909.301	4,973.466

Note: *p<0.05; **p<0.01; ***p<0.001

F General Linear Regression Output for Gender Tweets during COP21

Table F1 Results of general linear regression for gender tweets during COP21

	Dependent variable:		
	Ratio of gender tweets by CSO observers at COP21		
	(1)	**(2)**	**(3)**
Women	25.932***	22.172***	24.746***
	(3.958)	(4.202)	(4.596)
Multi-issue CSOs		−1.605	
		(1.954)	
Issue-specific CSOs		−3.196*	
		(1.888)	
Business/industry		−7.279***	−4.712
		(2.267)	(2.946)
University		−7.498***	−4.806*
		(2.025)	(2.719)
Other sectors		−3.424***	−0.837
		(1.849)	(2.629)
Operation scale	0.297	0.770	0.662
	(0.560)	(0.585)	(0.589)
Global North	1.639	2.444*	2.218
	(1.316)	(1.315)	(1.357)
Climate change			1.798
			(3.109)
Environmental			0.500
			(2.838)
Built environment			−2.099
			(3.981)
Conservation			0.641
			(4.129)
Finance/market mechanism			−3.449
			(4.591)
Food/agriculture			−4.440
			(4.049)

Table F1 (Cont.)

	Dependent variable:		
	Ratio of gender tweets by CSO observers at COP21		
	(1)	**(2)**	**(3)**
Forest			5.592
			(3.526)
Health			−1.867
			(4.591)
Legal			−0.982
			(6.245)
Rights/justice			0.768
			(3.392)
Sustainable development			−2.667
			(3.909)
Transport			−2.733
			(5.441)
Indigenous peoples			−3.310
			(3.912)
Youth/children			5.248
			(3.310)
Constant	3.884*** 6.400***		4.137
	(1.173)	(1.841)	(2.747)
Observations	786	786	786
Log likelihood	−3,129.913	−3,117.807	−3,110.719
Akaile Inf. Crit.	6,267.825	6,253.615	6,263.438

Note: *p<0.05; **p<0.01; ***p<0.001

G General Linear Regression Output for Indigenous Tweets during COP21

Table G1 Results of general linear regression for indigenous tweets during COP21

	Dependent variable:		
	Ratio of indigenous tweets by CSO observers at COP21		
	(1)	(2)	(3)
Indigenous CSOs	−1.225	0.286	−1.852
	(4.236)	(4.753)	(5.328)
Multi-issue CSOs		0.815	
		(2.805)	
Issue-specific CSOs		5.528**	
		(2.704)	
Business/industry		−3.162	−5.534
		(3.212)	(4.012)
University		−3.291	−5.712
		(2.848)	(3.702)
Other sectors		1.474	−0.890
		(2.654)	(3.580)
Operation scale	1.213	1.598**	1.618**
	(0.767)	(0.800)	(0.802)
Global North	−0.043	0.492	0.892
	(1.826)	(1.821)	(1.848)
Climate change			−1.625
			(4.234)
Environmental			−1.439
			(3.866)
Built environment			−1.281
			(5.422)
Conservation			12.134**
			(5.623)
Finance/market mechanism			−4.408
			(6.253)
Food/agriculture			6.264
			(5.515)

Table G1 (Cont.)

	Dependent variable:		
	Ratio of indigenous tweets by CSO observers at COP21		
	(1)	**(2)**	**(3)**
Forest			9.522**
			(4.803)
Health			10.320*
			(6.253)
Legal			−5.588
			(8.506)
Rights/justice			5.750
			(4.620)
Sustainable development			−5.256
			(5.323)
Transport			−6.885
			(7.411)
Women			−2.187
			(6.259)
Youth/children			−4.073
			(4.507)
Constant	7.094***	5.191*	7.192*
	(1.659)	(2.786)	(3.741)
Observations	786	786	786
Log likelihood	−3,376.861	−3,365.414	−3,353.524
Akaile Inf. Crit.	6,761.722	6,748.827	6,749.049

Note: *p<0.05; **p<0.01; ***p<0.001

H Interview Questions

1. Could you tell me your name, the organization that you are currently working with, and your position in the organization? Could you also give me a brief description of your current job?
2. Our research is interested in the strategies of CSOs at the UNFCCC, especially the framing and advocacy of the main issues your organization is concerned about. How would you describe your organization's main mission? Which climate-related issues does your organization prioritize?
3. What are your organization's main goals at the COP?
4. In order to achieve these goals, what types of strategies and activities does your organization take on?
5. Does your organization utilize social media platforms? If so, how does it use the platforms and for what purposes?
6. Besides your own organization, which other organizations would you consider playing a key role in the women's/indigenous advocacy and activism? What kind of role do they play?
7. (For generalist CSOs) We understand that your organization focuses on a range of climate issues such as ..., are there particular reasons it chooses to participate in the advocacy for women/indigenous people? How does the organization balance different issue areas?
8. How would you evaluate the outcome and effectiveness of your engagement at the COP? What are some of the biggest gains for your organization? What are the main obstacles?
9. How would you describe your feelings toward climate change? In your opinion, who are the victims and who should be responsible to address this global challenge?
10. What are the things that make your organization stand out from others working on climate change?
11. Do you find that there are too many organizations in the area you work on or do you think there should be more?

References

Abbott, Kenneth W., Jessica F. Green, and Robert O. Keohane. 2016. "Organizational ecology and institutional change in global governance." *International Organization* 70(2):247–277.

Albin, Cecilia. 1999. "Can NGOs enhance the effectiveness of international negotiation?" *International Negotiation* 4(3):371–387.

Aldrich, Howard E. and Jeffrey Pfeffer. 1976. "Environments of organizations." *Annual Review of Sociology* 2(1):79–105.

Allan, Jen Iris. 2020. *The new climate activism: NGO authority and participation in climate change governance*. University of Toronto Press.

Allan, Jen Iris and Jennifer Hadden. 2017. "Exploring the framing power of NGOs in global climate politics." *Environmental Politics* 26(4):600–620.

Alston, Margaret. 2015. *Women and climate change in Bangladesh*. Routledge.

Bäckstrand, Karin and Eva Lövbrand. 2006. "Planting trees to mitigate climate change: Contested discourses of ecological modernization, green governmentality and civic environmentalism." *Global Environmental Politics* 6(1):50–75.

Bäckstrand, Karin and Eva Lövbrand. 2019. "The road to Paris: Contending climate governance discourses in the post-Copenhagen era." *Journal of Environmental Policy & Planning* 21(5):519–532.

Bagozzi, Benjamin E. 2015. "The multifaceted nature of global climate change negotiations." *The Review of International Organizations* 10:439–464.

Balboa, Cristina M. 2018. *The paradox of scale: How NGOs build, maintain, and lose authority in environmental governance*. MIT Press.

Baldassarri, Delia and Peter Bearman. 2007. "Dynamics of political polarization." *American Sociological Review* 72(5):784–811.

Barnett, Michael and Peter Walker. 2015. "Regime change for humanitarian aid." *Foreign Affairs* 94:130.

Baum, Joel A. C. and Christine Oliver. 1991. "Institutional linkages and organizational mortality." *Administrative Science Quarterly* 36(2):187–218.

Belfer, Ella, James D. Ford, Michelle Maillet, Malcolm Araos, and Melanie Flynn. 2019. "Pursuing an indigenous platform: Exploring opportunities and constraints for indigenous participation in the UNFCCC." *Global Environmental Politics* 19(1):12–33.

Benford, Robert D. and David A. Snow. 2000. "Framing processes and social movements: An overview and assessment." *Annual Review of Sociology* 26(1):611–639.

Bernstein, Steven. 2001. *The compromise of liberal environmentalism*. Columbia University Press.

Betsill, Michele M. and Elisabeth Corell, eds. 2008. *NGO diplomacy: The influence of nongovernmental organizations in international environmental negotiations*. MIT Press.

Bexell, Magdalena, Jonas Tallberg, and Anders Uhlin. 2010. "Democracy in global governance: The promises and pitfalls of transnational actors." *Global Governance* 16(1):81–101.

Blei, David M., Andrew Y. Ng, and Michael I. Jordan. 2003. "Latent dirichlet allocation." *Journal of Machine Learning Research* 3(Jan):993–1022.

Bob, Clifford. 2005. *The marketing of rebellion: Insurgents, media, and international activism*. Cambridge University Press.

Bode, Leticia, Alexander Hanna, Junghwan Yang, and Dhavan V. Shah. 2015. "Candidate networks, citizen clusters, and political expression: Strategic hashtag use in the 2010 midterms." *The ANNALS of the American Academy of Political and Social Science* 659(1):149–165.

Bond, Patrick and Michael K. Dorsey. 2010. "Anatomies of environmental knowledge and resistance: Diverse climate justice movements and waning eco-neoliberalism." *Journal of Australian Political Economy* 66:286–316.

Boscarino, Jessica E. 2016. "Setting the record straight: Frame contestation as an advocacy tactic." *Policy Studies Journal* 44(3):280–308.

Boyd, Danah, Scott Golder, and Gilad Lotan. 2010. "Tweet, tweet, retweet: Conversational aspects of retweeting on twitter." *43rd Hawaii international conference on system sciences*. IEEE, pp. 1–10.

Buckingham, Susan. 2010. "Call in the women." *Nature* 468(7323):502.

Bush, Sarah Sunn and Jennifer Hadden. 2019. "Density and decline in the founding of international NGOs in the United States." *International Studies Quarterly* 63(4):1133–1146.

Cabré, Miquel Muñoz. 2011. "Issue-linkages to climate change measured through NGO participation in the UNFCCC." *Global Environmental Politics* 11(3):10–22.

Carpenter, Charli. 2007. "Setting the advocacy agenda: Theorizing issue emergence and nonemergence in transnational advocacy networks." *International Studies Quarterly* 51(1):99–120.

Carroll, Glenn R. 1985. "Concentration and specialization: Dynamics of niche width in populations of organizations." *American Journal of Sociology* 90(6):1262–1283.

Charmaz, Kathy, Liska Belgrave, Jaber F. Gubrium, and James A. Holstein, eds. 2012. *The SAGE handbook of interview research: The complexity of the craft*. Sage.

Ciplet, David. 2014. "Contesting climate injustice: Transnational advocacy network struggles for rights in UN climate politics." *Global Environmental Politics* 14(4):75–96.

Claeys, Priscilla and Deborah Delgado Pugley. 2017. "Peasant and indigenous transnational social movements engaging with climate justice." *Canadian Journal of Development Studies/Revue canadienne d'études du développement* 38(3):325–340.

Comberti, Claudia, Thomas F. Thornton, Michaela Korodimou, Meghan Shea, and Kimaren Ole Riamit. 2019. "Adaptation and resilience at the margins: Addressing indigenous peoples' marginalization at international climate negotiations." *Environment: Science and Policy for Sustainable Development* 61(2):14–30.

Cooley, Alexander and James Ron. 2002. "The NGO scramble: Organizational insecurity and the political economy of transnational action." *International Security* 27(1):5–39.

De Moor, Joost. 2018. "The 'efficacy dilemma' of transnational climate activism: The case of COP21." *Environmental Politics* 27(6):1079–1100.

de Moor, Joost. 2022. "Alternative globalities? Climatization processes and the climate movement beyond COPs." In Stefan Aykut, and Lucile Maertens, eds., *The climatization of global politics*. Springer pp. 83–100.

Dietrich, Simone and Amanda Murdie. 2017. "Human rights shaming through INGOs and foreign aid delivery." *The Review of International Organizations* 12:95–120.

Disch, Lisa. 2012. "Democratic representation and the constituency paradox." *Perspectives on Politics* 10(3):599–616.

Dowling, Robyn. 2010. "Geographies of identity: Climate change, governmentality and activism." *Progress in Human Geography* 34(4):488–495.

Eilstrup-Sangiovanni, Mette. 2019. "Competition and strategic differentiation among transnational advocacy groups." *Interest Groups & Advocacy* 8(3):376–406.

Eilstrup-Sangiovanni, Mette. 2021. "What kills international organisations? When and why international organisations terminate." *European Journal of International Relations* 27(1):281–310.

Epstein, Charlotte. 2008. *The power of words in international relations: Birth of an anti-whaling discourse*. MIT Press.

Finnemore, Martha and Kathryn Sikkink. 1998. "International norm dynamics and political change." *International Organization* 52(4):887–917.

Finnemore, Martha and Kathryn Sikkink. 2001. "Taking stock: The constructivist research program in international relations and comparative politics." *Annual Review of Political Science* 4(1):391–416.

Ford, James, Michelle Maillet, Vincent Pouliot, Thomas Meredith, Alicia Cavanaugh, and IHACC Research Team. 2016. "Adaptation and indigenous peoples in the United Nations framework convention on climate change." *Climatic Change* 139:429–443.

Fox, Jonathan A. and L David Brown. 1998. *The struggle for accountability: The World Bank, NGOs, and grassroots movements*. MIT press.

Freeman, John and Michael T. Hannan. 1983. "Niche width and the dynamics of organizational populations." *American Journal of Sociology* 88(6):1116–1145.

Friedman, Elisabeth Jay. 2003. "Gendering the agenda: The impact of the transnational women's rights movement at the UN conferences of the 1990s." *Women's Studies International Forum* 26(4):313–331.

Gamson, William A., Andre Modigliani. 1987. "The changing culture of affirmative action." *Research in Political Sociology* 3(1):137–177.

Gardner, William, Edward P. Mulvey, and Esther C. Shaw. 1995. "Regression analyses of counts and rates: Poisson, overdispersed Poisson, and negative binomial models." *Psychological Bulletin* 118(3):392.

Gent, Stephen E., Mark J. C. Crescenzi, Elizabeth J. Menninga, and Lindsay Reid. 2015. "The reputation trap of NGO accountability." *International Theory* 7(3):426–463.

Glaser, Barney and Anselm Strauss. 1967. *The discovery of grounded theory: Strategies for qualitative research*. Sociology Press.

Goffman, Erving. 1974. *Frame analysis: An essay on the organization of experience*. Harvard University Press.

Hadden, Jennifer. 2015. *Networks in contention*. Cambridge University Press.

Hafner-Burton, Emilie M. 2008. "Sticks and stones: Naming and shaming the human rights enforcement problem." *International Organization* 62(4): 689–716.

Hafner-Burton, Emilie and Mark A. Pollack. 2002. "Mainstreaming gender in global governance." *European Journal of International Relations* 8(3): 339–373.

Hannan, Michael T. and John Freeman. 1977. "The population ecology of organizations." *American Journal of Sociology* 82(5):929–964.

Hannan, Michael T. and John Freeman. 1989. *Organizational ecology*. Harvard University Press.

Hastie, Trevor, Robert Tibshirani, Jerome H. Friedman and Jerome H. Friedman. 2009. *The elements of statistical learning: Data mining, inference, and prediction*. Vol. 2 Springer.

Hemphill, Libby and Andrew J. Roback. 2014. "Tweet acts: How constituents lobby congress via Twitter." In *Proceedings of the 17th ACM conference on computer supported cooperative work & social computing*. pp. 1200–1210.

Hendrix, Cullen S. and Wendy H. Wong. 2013. "When is the pen truly mighty? Regime type and the efficacy of naming and shaming in curbing human rights abuses." *British Journal of Political Science* 43(3):651–672.

Hildebrandt, Timothy and Lynette J. Chua. 2017. "Negotiating in/visibility: The political economy of lesbian activism and rights advocacy." *Development and Change* 48(4):639–662.

Hjerpe, Mattias and Katarina Buhr. 2014. "Frames of climate change in side events from Kyoto to Durban." *Global Environmental Politics* 14(2): 102–121.

Hopke, Jill E. and Luis E. Hestres. 2018. "Visualizing the Paris climate talks on Twitter: Media and climate stakeholder visual social media during COP21." *Social Media+ Society* 4(3):2056305118782687.

Indigenous Peoples of North America. 1998. "A call to action: The Albuquerque declaration."

Interview 1701. 2017. GenderCC interviewed by Author B, Bonn, Germany.

Interview 1702. 2017. GenderCC interviewed by Author B, Bonn, Germany.

Interview 1703. 2017. Women's Environment and Develop Organization (WEDO) interviewed by Author B, Bonn, Germany.

Interview 1801. 2018. Plan International interviewed by Author B, Katowice, Poland.

Interview 1802. 2018. Asian-Pacific Resource and Research Center for Women (ARROW) interviewed by Author B, Katowice, Poland.

Interview 1803. 2018. Centre for 21st Century Issues interviewed by Author B, Katowice, Poland.

Interview 1805. 2018. December 13. Interviewed by Author B, Katowice, Poland.

Interview 2202. 2022. ADRA interviewed by Author A, Sharm El-Sheikh, Egypt.

Interview 2203. 2022. Care International interviewed by Author A, Sharm El-Sheikh, Egypt.

Interview 2204. 2022. Centre for Feminist Foreign Policy interviewed by Author A, Sharm El-Sheikh, Egypt.

Interview 2205. 2022. Conservation International interviewed by Author A, Sharm El-Sheikh, Egypt.

Interview 2206. 2022. Cultural Survival interviewed by Author A, Sharm El-Sheikh, Egypt.

Interview 2207. 2022. Forest Trend interviewed by Author A, Sharm El-Sheikh, Egypt.

Interview 2208. 2022. IBON interviewed by Author A, Sharm El-Sheikh, Egypt.

Interview 2209. 2022. ICCC interviewed by Author A, Sharm El-Sheikh, Egypt.

Interview 2211. 2022. NEFIN interviewed by Author A, Sharm El-Sheikh, Egypt.

Interview 2212. 2022. Plan International Canada interviewed by Author A, Sharm El-Sheikh, Egypt.

Interview 2213. 2022. Agroforestry interviewed by Author A, Sharm El-Sheikh, Egypt.

Interview 2214. 2022. WWF interviewed by Author A, Sharm El-Sheikh, Egypt.

Jinnah, Sikina. 2011. "Climate change bandwagoning: The impacts of strategic linkages on regime design, maintenance, and death." *Global Environmental Politics* 11(3):1–9.

Keck, Margaret E. and Kathryn Sikkink. 1998. *Activists beyond borders*. Cornell University Press.

Kruskal, Joseph B. and Myron Wish. 1978. *Multidimensional scaling*. Vol. 11 Sage.

Kuyper, Jonathan W. and Karin Bäckstrand. 2016. "Accountability and representation: Nonstate actors in UN climate diplomacy." *Global Environmental Politics* 16(2):61–81.

Lake, David A. 2021. "The organizational ecology of global governance." *European Journal of International Relations* 27(2):345–368.

López-Rivera, Andrés. 2023. "Diversifying Boundary Organizations: The Making of a Global Platform for Indigenous (and Local) Knowledge in the UNFCCC." *Global Environmental Politics* 23(4):52–72.

Maldonado, Julie, T. M. Bull Bennett, Karletta Chief, et al. 2016. "Engagement with indigenous peoples and honoring traditional knowledge systems." *The US National Climate Assessment: Innovations in Science and Engagement* 135:111–126.

Mansbridge, Jane. 2003. "Rethinking representation." *American Political Science Review* 97(4):515–528.

Merry, Melissa K. 2013. "Tweeting for a cause: Microblogging and environmental advocacy." *Policy & Internet* 5(3):304–327.

Minkoff, Debra C. 1997. "The sequencing of social movements." *American Sociological Review* 62(5):779–799.

Morin, Jean-Frédéric. 2020. "Concentration despite competition: The organizational ecology of technical assistance providers." *The Review of International Organizations* 15(1):75–107.

Murdie, Amanda. 2014. "The ties that bind: A network analysis of human rights international nongovernmental organizations." *British Journal of Political Science* 44(1):1–27.

Nagel, Joane. 2015. *Gender and climate change: Impacts, science, policy*. Routledge.

Nisbet, Matthew C. and Dietram A. Scheufele. 2009. "What's next for science communication? Promising directions and lingering distractions." *American Journal of Botany* 96(10):1767–1778.

North, Douglass C. 1990. *Institutions, institutional change and economic performance*. Cambridge University Press.

Olson, Mancur. 1971. *The logic of collective action: Public goods and the theory of groups, with a new preface and appendix*. Vol. 124 Harvard University Press.

O'Neill, Saffron J., Mike Hulme, John Turnpenny, and James A. Screen. 2010. "Disciplines, geography, and gender in the framing of climate change." *Bulletin of the American Meteorological Society* 91(8):997–1002.

Pallas, Christopher L. and Lan Nguyen. 2018. "Transnational advocacy without northern NGO partners: Vietnamese NGOs in the HIV/AIDS sector." *Nonprofit and Voluntary Sector Quarterly* 47(4_suppl):159S–176S.

Pierson, Paul. 2000. "Increasing returns, path dependence, and the study of politics." *American Political Science Review* 94(2):251–267.

Prakash, Aseem and Mary Kay Gugerty, eds. 2010. *Advocacy organizations and collective action*. Cambridge University Press.

Price, Richard. 1998. "Reversing the gun sights: Transnational civil society targets land mines." *International Organization* 52(3):613–644.

Raffel, Sara. 2016. "Climate communication and the exclusion of indigenous knowledge." In *2016 IEEE International Professional Communication Conference (IPCC)*. pp. 1–5.

Ramesh, Arti, Dan Goldwasser, Bert Huang, Hal Daumé III, and Lise Getoor. 2014. "Understanding MOOC discussion forums using seeded LDA." In *Proceedings of the 9th workshop on innovative use of NLP for building educational applications*. pp. 28–33.

Rashidi, Pedram and Kristen Lyons. 2021. "Democratizing global climate governance? The case of indigenous representation in the Intergovernmental Panel on Climate Change (IPCC)." *Globalizations* 20(8):1–16.

Saward, Michael. 2006. "The representative claim." *Contemporary Political Theory* 5:297–318.

Schapper, Andrea, Linda Wallbott, and Katharina Glaab. 2023. "The climate justice community: Theoretical radicals and practical pragmatists?" *Global Society* 37(3):397–419.

Schroeder, Heike. 2010. "Agency in international climate negotiations: The case of indigenous peoples and avoided deforestation." *International Environmental Agreements: Politics, Law and Economics* 10:317–332.

Shawki, Noha. 2011. "Organizational structure and strength and transnational campaign outcomes: A comparison of two transnational advocacy networks." *Global Networks* 11(1):97–117.

Shawoo, Zoha and Thomas F. Thornton. 2019. "The UN local communities and Indigenous peoples' platform: A traditional ecological knowledge-based evaluation." *Wiley Interdisciplinary Reviews: Climate Change* 10(3):e575.

Shearer, Elisa, Michael Barthel, Jeffrey Gottfried, and Amy Mitchell. 2015. "The evolving role of news on Twitter and Facebook." *Pew Research Center*.

Shibaike, Takumi. 2022. "Small NGOs and agenda-setting in global conservation governance: The case of pangolin conservation." *Global Environmental Politics* 22(2):45–69.

Shibaike, Takumi. 2023. "The power of specialization: NGO advocacy in global conservation governance." *International Studies Quarterly* 67(2):sqad023.

Smith, Tonya, Janette Bulkan, Hisham Zerriffi, and James Tansey. 2019. "Indigenous peoples, local communities, and payments for ecosystem services." *The Canadian Geographer/Le Géographe Canadien* 63(4):616–630.

Snow, David and Scott Byrd. 2007. "Ideology, framing processes, and Islamic terrorist movements." *Mobilization: An International Quarterly* 12(2): 119–136.

Snow, David A., E. Burke Rochford Jr., Steven K. Worden, and Robert D. Benford. 1986. "Frame alignment processes, micromobilization, and movement participation." *American Sociological Review* 51(4):464–481.

Steffek, Jens. 2013. "Explaining cooperation between IGOs and NGOs – push factors, pull factors, and the policy cycle." *Review of International Studies* 39(4):993–1013.

Steffek, Jens, Claudia Kissling, and Patrizia Nanz. 2007. *Civil society participation in European and global governance: A cure for the democratic deficit?* Springer.

Stroup, Sarah S. and Wendy H. Wong. 2017. *The authority trap: Strategic choices of international NGOs*. Cornell University Press.

Tallberg, Jonas, Thomas Sommerer, and Theresa Squatrito. 2013. *The opening up of international organizations*. Cambridge University Press.

Tarrow, Sidney. 2005. *The new transnational activism*. Cambridge University Press.

Tormos-Aponte, Fernando. 2021. "The influence of indigenous peoples in global climate governance." *Current Opinion in Environmental Sustainability* 52:125–131.

UNFCCC. 2012. "Promoting gender equality and improving the participation of women in UNFCCC negotiations and in the representation of Parties in bodies established pursuant to the Convention or the Kyoto Protocol."

UNFCCC. 2017. "Report of the conference of the parties on its twenty-third session."

Vanhala, Lisa and Cecilie Hestbaek. 2016. "Framing climate change loss and damage in UNFCCC negotiations." *Global Environmental Politics* 16(4):111–129.

Watanabe, Kohei and Yuan Zhou. 2022. "Theory-driven analysis of large corpora: Semisupervised topic classification of the UN speeches." *Social Science Computer Review* 40(2):346–366.

Willetts, Peter. 2001. "Transnational actors and international organizations in global politics." *The Globalization of World Politics* 2:356–383.

Wong, Wendy, Ron Levi, and Julia Deutsch. 2017. "The Ford Foundation." In *Professional networks in transnational governance*, ed. Leonard Seabrooke and Lasse Folke Henriksen. Cambridge University Press Cambridge pp. 82–100.

Yu, Bei, Stefan Kaufmann, and Daniel Diermeier. 2008. "Classifying party affiliation from political speech." *Journal of Information Technology & Politics* 5(1):33–48.

Zeng, Fanxu, Jia Dai, and Jeffrey Javed. 2019. "Frame alignment and environmental advocacy: The influence of NGO strategies on policy outcomes in China." *Environmental Politics* 28(4):747–770.

Zhao, Bi. 2023. "Granting legitimacy from non-state actor deliberation: An example of women's groups at the United Nations Framework Convention on Climate Change." *Environmental Policy and Governance* 34(3): 236–255.

Zurba, Melanie and Anastasia Papadopoulos. 2023. "Indigenous participation and the incorporation of indigenous knowledge and perspectives in global environmental governance forums: A systematic review." *Environmental Management* 72: 84–99.

Cambridge Elements

Organizational Response to Climate Change

Matthew Potoski

UC Santa Barbara

Matthew Potoski is a Professor at UCSB's Bren School of Environmental Science and Management. He currently teaches courses on corporate environmental management, and his research focuses on management, voluntary environmental programs, and public policy. His research has appeared in business journals such as *Strategic Management Journal, Business Strategy and the Environment*, and the *Journal of Cleaner Production*, as well as public policy and management journals such as *Public Administration Review* and the *Journal of Policy Analysis and Management*. He coauthored *The Voluntary Environmentalists* (Cambridge, 2006) and *Complex Contracting* (Cambridge, 2014; the winner of the 2014 Best Book Award, American Society for Public Administration, Section on Public Administration Research) and was coeditor of *Voluntary Programs* (MIT, 2009). Professor Potoski is currently coeditor of the *Journal of Policy Analysis and Management* and the *International Public Management Journal*.

About the Series

How are governments, businesses, and nonprofits responding to the climate challenge in terms of what they do, how they function, and how they govern themselves? This series seeks to understand why and how they make these choices and with what consequence for the organization and the eco-system within which it functions.

Cambridge Elements

Organizational Response to Climate Change

Elements in the Series

Explaining Transformative Change in ASEAN and EU Climate Policy: Multilevel Problems, Policies and Politics
Charanpal Bal, David Coen, Julia Kreienkamp, Paramitaningrum, and Tom Pegram

Fighting Climate Change through Shaming
Sharon Yadin

Governing Sea Level Rise in a Polycentric System: Easier Said than Done
Francesca Vantaggiato and Mark Lubell

Inside the IPCC: How Assessment Practices Shape Climate Knowledge
Jessica O'Reilly, Mark Vardy, Kari De Pryck, and Marcela da S. Feital Benedetti

Climate Activism, Digital Technologies, and Organizational Change
Mette Eilstrup-Sangiovanni and Nina Hall

Who Tells Your Story? Women and Indigenous Peoples Advocacy at the UNFCCC
Takumi Shibaike and Bi Zhao

A full series listing is available at: www.cambridge.org/ORCC

Printed in the United States
by Baker & Taylor Publisher Services